Astrobiology

Monica M. Grady

Smithsonian Institution Press, Washington, D.C.
in association with The Natural History Museum, London

To Ian and Jack, with love.

Published in the United States of America
by the Smithsonian Institution Press in association with
The Natural History Museum
London
Cromwell Road
London SW7 5BD
United Kingdom

Library of Congress Cataloging-in-Publication Data
Grady, M. M. (Monica M.)
    Astrobiology / Monica Grady.
        p.   cm.
    Includes bibliographical references and index.
    ISBN 1-56098-849-5 (alk. paper)
    1. Life on other planets. 2. Exobiology. I. Title.

QB54.G73   2001
576.8'39—dc21   00-046321

Manufactured in Singapore, not at government expense
08 07 06 05 04 03 02 01 5 4 3 2 1

DISTRIBUTION

United States, Canada, Central and South America
and the Caribbean
Smithsonian Institution Press
470 L'Enfant Plaza
Washington, D.C. 20560-0950
USA

Australia and New Zealand
CSIRO Publishing
PO Box 1139
Collingwood, Victoria 3066
Australia

UK and rest of the world
Plymbridge Distributors Ltd.
Plymbridge House, Estover Road
Plymouth, Devon PL6 7PY
UK

Edited by Celia Coyne
Designed by Mercer Design
Reproduction and printing by Craft Print, Singapore

On the front cover: Multiple generations of stars
in the Tarantula nebula. Inset: Pathfinder Rover.
On the back cover and title page: Mars

Contents

Preface

The question of whether we are alone in our Solar System, Galaxy or Universe has fascinated humanity since the earliest of times. Stories of mysterious beings from the sky permeate the mythology of many cultures, and make a regular appearance in fiction. The number of 'UFO' sightings also continues to rise, even though fairly mundane explanations account for practically all observations. Yet aside from the myths and sensationalism, both the search for extraterrestrial intelligence (SETI) and the study of astrobiology (or exobiology) have become widely accepted as valid and important areas of research. This book focuses on astrobiology, the search for life outside our planet. By taking a broad view of the origin and evolution of life on Earth, astrobiology considers the possibilities for similar developments in other planetary environments within the Solar System and beyond. Starting with the Big Bang and progressing to the development of humanity on Earth, astrobiology encompasses cosmologists, astrophysicists, astronomers, planetary scientists, palaeontologists, chemists, biochemists, biologists, geneticists and anthropologists, not forgetting the philosophers and theologians who are captivated by the range and depth of the subject matter. In order to understand the variety of life on Earth, we must understand something of the Earth itself: its place within the Solar System; the place of its star, the Sun, within our Galaxy (the Milky Way); and, ultimately, the history of our Galaxy within the Universe. So we start the search for life at the beginning – with the Big Bang, when the Universe came into being.

Author

Dr Monica Grady is Head of the Petrology and Meteoritics Division in the Department of Mineralogy at The Natural History Museum, and Honorary Reader in Geological Sciences at University College, London. Monica received an honours degree in Chemistry and Geology from the University of Durham in 1979, then went on to complete a PhD on carbon in stony meteorites at the University of Cambridge in 1982. Since then, Monica has continued to specialise in the study of meteorites, and carried out this research at Cambridge, then the Open University in Milton Keynes, prior to joining The Natural History Museum in 1991. Her particular research interests are in the fields of carbon and nitrogen stable isotope geochemistry of Martian meteorites, interstellar components in meteorites, micrometeorites, and also in astrobiology and the possibilities of life elsewhere in the cosmos. Asteroid (4731) was named 'Monicagrady' in her honour.

The origins of life

The search for the origins of life must necessarily go back as far as time itself – back to the point when time and space began, and all matter and energy were created. The Big Bang is the name that has been given to the process that formed the Universe. There is no 'before' the Big Bang, just as there is no 'outside' the Universe. *That doesn't make sense.*

The Big Bang

The Big Bang occurred approximately 15 billion years ago, when a point with unimaginable temperature, pressure and density exploded. Cosmology is the study of the origin and evolution of the Universe, and cosmologists can speculate on processes that took place as far back as a tiny fraction of a second after the Big Bang (see p. 6).

Within about a minute of the Big Bang the Universe had expanded by a million times, and it is still expanding today. As the Universe rapidly expanded it also cooled down. After 300,000 years or so, tiny particles (protons, neutrons and electrons) began to form atoms, mainly of hydrogen and helium. After about two billion years, matter started to clump together to form the first galaxies and stars, and after around 10 billion years, our star, the Sun, with its accompanying planets, asteroids and comets, was born from an interstellar cloud of gas and dust.

The elements hydrogen and helium that were produced shortly after the Big Bang are the main constituents of stars and galaxies.

But planets (and lifeforms) are composed of additional elements such as carbon, oxygen, silicon, iron and magnesium. These elements were not made in the Big Bang, but in reactions taking place inside stars (see p. 7) and then ejected into space when stars exploded. In this way the interstellar medium is continually enriched as elements are recycled through generations of star formation.

ABOVE **Part of the Hubble Deep Field showing hundreds of different galaxies, some of which may have formed less than 1 billion years after the Big Bang.**

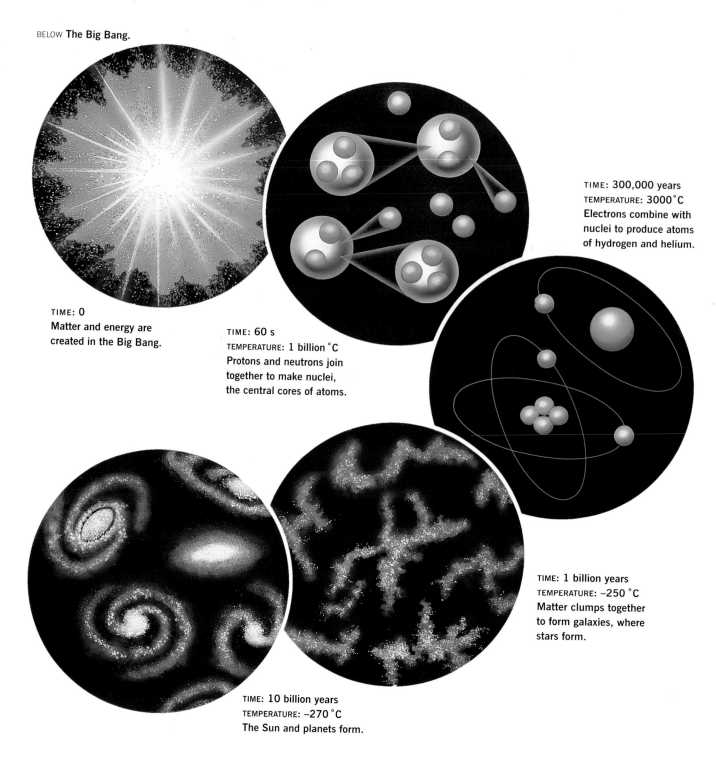

BELOW **The Big Bang.**

TIME: 0
Matter and energy are
created in the Big Bang.

TIME: 60 s
TEMPERATURE: 1 billion °C
Protons and neutrons join
together to make nuclei,
the central cores of atoms.

TIME: 300,000 years
TEMPERATURE: 3000 °C
Electrons combine with
nuclei to produce atoms
of hydrogen and helium.

TIME: 1 billion years
TEMPERATURE: −250 °C
Matter clumps together
to form galaxies, where
stars form.

TIME: 10 billion years
TEMPERATURE: −270 °C
The Sun and planets form.

LEFT **The Helix nebula is a planetary nebula. A cool giant star exploded and has thrown gas and dust back out into space. Only a small white dwarf star remains at the centre.**

The most ancient stars (10–12 billion years old) contain extremely low levels of any elements other than hydrogen and helium, indicating that they formed from the most primordial of material. Younger stars like the Sun, with an age of approximately five billion years, contain significant amounts of the heavier elements (carbon, oxygen, silicon, etc.). This is because by the time such stars were born, the Universe had already experienced 10 billion years of element production. Matter drifts through space as clouds of gas and dust. The interstellar clouds are mostly hydrogen, with carbon monoxide plus a variety of additional carbon-based (organic) molecules, grains of silicate dust and ices. It is the organic molecules that form the building blocks of life. But how did this jumble of dust, ices and organics become the Sun and Solar System and how was life initiated on Earth?

LEFT **The Crab nebula is a supernova remnant from a star that exploded in 1054. Hot gas is expanding from the central pulsar at a speed of about 1500 kms^{-1} (around 3.4 million mph).**

BELOW **Diagram to show the structure of the Milky Way Galaxy, and the position of the Earth relative to the galactic centre.**

Formation of the Solar System

Our Sun and its planetary system is located approximately two-thirds of the way out from the centre, between two of the arms of the Milky Way spiral galaxy. The Sun is sometimes described as an undistinguished star in an undistinguished neighbourhood – it is a main sequence star surrounded by its family of planets and their satellites, plus asteroids and comets. The first stage in the history of the Solar System was when an interstellar cloud collapsed to form a 'protoplanetary disk' (the solar nebula). Astronomers do not know what triggered the cloud collapse and have suggested several possibilities, such as a shockwave from a nearby supernova. As the cloud collapsed,

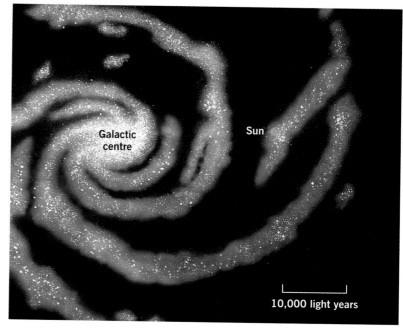

Galactic centre

Sun

10,000 light years

Forming elements inside stars

Elements that are heavier than hydrogen and helium are made by different types of nuclear reactions in stars, collectively termed 'nucleosynthesis'. The enormous amount of heat and light energy emitted by a star is the result of nuclear fusion at its core. As main sequence stars (like the Sun) burn hydrogen, the hydrogen atoms fuse together to produce first deuterium, then helium, with an accompanying release of energy. Eventually, when all the hydrogen is exhausted, helium burning begins, fusing three helium nuclei together to produce one carbon atom. Successive cycles of nucleosynthesis continue by fusion of nuclei, forming elements with greater and greater masses – carbon, oxygen, neon, magnesium, silicon. Finally, iron is produced and fusion stops – above this mass, fusion of nuclei requires energy to be taken in, rather than given out.

What happens next depends on the size and temperature of the star. A massive star (more than about 10 times the mass of the Sun) collapses inwards, then explodes as a supernova, radiating huge amounts of energy and matter. Remnant material is left behind either as a neutron star, or as a black hole, depending on the initial mass of the star. The intense energy of the supernova explosion fuels nuclear reactions between neutrons released from the exploding star and the accelerating atomic nuclei ejected from the star's interior. In cooler stars such as Red Giants, reactions between atoms and neutrons occur during cycling of material between different layers within the star. The final stage of a Red Giant is when it casts off a shell of material (a planetary nebula) into space, allowing heavy elements to be ejected back into the interstellar medium; all that is left behind is a white dwarf star.

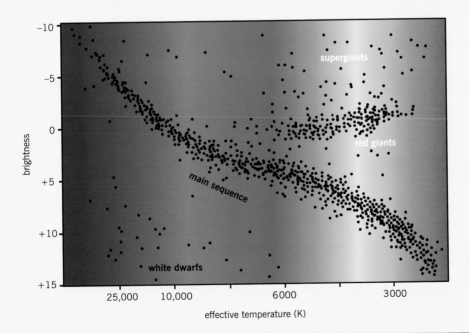

LEFT **The evolution of a star can be plotted on a Herzsprung-Russell diagram, which shows how the temperature and brightness (luminosity) of stars vary. Temperature ranges from hot, blue stars to cool, red stars.**

dust grains fell towards the centre of the nebula, clumping together to form increasingly large bodies, and eventually forming into planets. The growth of planet-sized bodies from micron-sized dust grains is controlled by several factors, such as the nature of the grains (fluffy or compact) and the degree of turbulence within the nebula.

LEFT **Formation of the Solar System**

1 **A turbulent cloud of interstellar dust and gas collapses.**

2 **The dust and gas cloud forms a spinning disk.**

3 **The temperature and pressure of the disk increases towards the middle. Eventually, the temperature is hot enough for fusion of hydrogen into helium to begin. The central star is born.**

4 **The remaining dust and gas clump together, gradually sweeping up all the debris into planets.**

5 **The Solar System now has 9 planets plus their satellites (or moons), asteroids, comets and Kuiper Belt objects.**

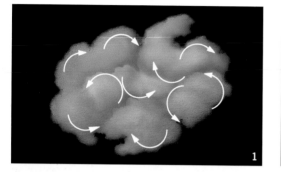
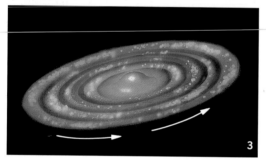
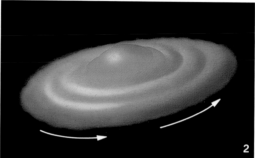
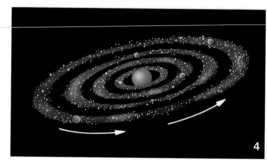
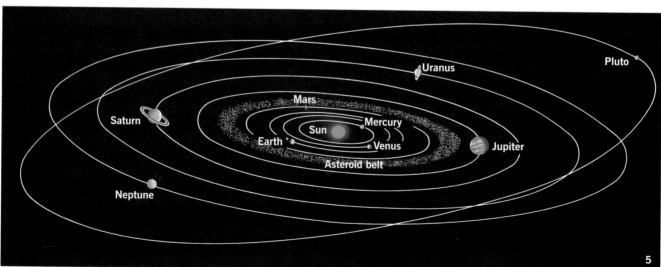

The collapse of the cloud and subsequent accretion of material must have taken place over a relatively short time interval, around 1–3 million years, as deduced by measurements on meteorites.

Rocky planets and asteroids

The inner, rocky planets (Mercury, Venus, Earth and Mars) formed sufficiently close to the Sun that the most volatile elements (such as hydrogen and helium) were lost through vaporisation, and water did not form ice. Farther out, marking the boundary between the rocky planets and the outer gas giants, is the Asteroid Belt, a swarm of several thousand small rocky planets (planetesimals). Asteroids are not the remains of a planet that formed and then broke up – rather, they are the leftovers from the planet-building process. The gravitational pull of Jupiter prevented the planetesimals from collecting into one body.

The largest asteroid, Ceres, is about 1000 km (620 miles) across, although most asteroids are just a few kilometres across. They are grouped into different types, depending on their composition. Some are predominantly composed of iron, whereas others are mostly stony; one group is very rich in carbon.

Although the asteroids are in stable orbits around the Sun, collisions between asteroids occur, as can be surmised from their highly cratered surfaces. Fragments of asteroids ejected from the Asteroid Belt towards the Sun may fall on to the inner planets. On Earth, these fragments are known as meteorites (see pp, 12–13). In order to understand how and why life has evolved on the Earth, and whether

ABOVE **Asteroid 433 Eros, photographed from an altitude of 200 km (140 miles) by the spacecraft NEAR-Shoemaker, February 2000. It is about 33 km long by 13 km wide (21 x 8 miles), and is a stony asteroid.**

it might exist in other places within the Solar System, it is necessary to know what materials the Earth formed from and how they have been changed through time. Meteorites are the only physical materials available on the Earth that allow direct study of the original dust from which the Earth, and the Solar System, formed.

The gas giants and beyond

The sequence of events that produced the innermost rocky planets also produced planets of an entirely different nature: the gas giants Jupiter, Saturn, Uranus and Neptune.

Meteorites

Meteorites are pieces of rock and metal that fall to the Earth. Almost all are fragments from asteroids. Study of meteorites allows a more complete understanding of the processes undergone by the material that resulted in the Earth of today. Meteorites can be divided into two main types, according to whether or not they have ever been molten.

Unmelted meteorites (or chondrites)

These are the most primitive of all meteorites. Chondrites are made from the same dust that made the Earth, but come from parent asteroids that have experienced only slight changes to their composition since their formation early in the history of the Solar System. Chondrites, therefore, are the direct ancestors of the Earth, and are studied to gain a better understanding of how the Earth has evolved to its present state. Chondrites are stone meteorites composed of calcium, aluminium-rich inclusions (CAIs), chondrules, metal and sulphides in a matrix of broken chondrules and minerals formed by the alteration of silicate minerals by water. CAIs are up to 1 cm (0.4 in) in size and include the minerals spinel, hibonite and melilite. The CAIs are the oldest of all planetary materials, and formed very soon after the initial collapse of the solar nebula. Chondrules are small, millimetre-sized spherical formations of the silicates olivine, pyroxene and plagioclase, and formed two or three million years after the CAIs. Study of CAIs and chondrules inform us about the time it has taken for dust to clump together to form parent bodies. This, in turn, allows us to estimate how much fuel (in the form of radioactive elements) was present in the solar nebula when the Earth aggregated, which in turn allows calculation of the timing of when the Earth's core formed.

One sub-group of chondrites contains up to 6% by weight of carbon, mainly as organic compounds, including the biologically significant amino and carboxylic acids. The presence of these molecules in meteorites is direct evidence that asteroids might have contributed some of the building blocks of life to the Earth during bombardment. Chondrites also contain very small (less than 1 micron in size), rare grains of diamond, graphite and carborundum (silicon carbide). These tiny crystals have come from stars and supernovae that existed before the Sun was formed.

FAR LEFT **A chondrite (15 cm across); meteorites like this are the compressed remnants of the dust from which the Solar System formed.**

LEFT **A broken fragment (5 cm across) from the Allende chondrite showing irregularly shaped CAIs, plus chondrules, the oldest solids to form from the solar nebula.**

By studying the composition and chemistry of these exotic grains, we can learn about the Sun's neighbours, and how the local galaxy has changed with time.

Melted meteorites

These come from asteroids that were so hot that they melted. Some are iron (with 5–15% by weight of nickel), some stone (made of the same minerals as chondrites), and the smallest group of all, the stony-irons, are an approximately equal mixture of stone and iron. Melted meteorites are slightly younger than the chondrites, and are much less primitive, having suffered extensive heating that separated metal from rock on their parent bodies. Study of melted meteorites can help to explain the processes involved in the formation of planets, especially core formation.

Non-asteroidal meteorites

Although almost all meteorites come from the Asteroid Belt, there are two groups of melted stony meteorites that have different origins.

Lunar meteorites: There are currently 18 known lunar meteorites, 15 of which were collected from Antarctica. The lunar meteorites are almost identical in composition, texture and chemistry to the lunar soils and rocks brought back from the Moon to the Earth by the Apollo astronauts.

Martian meteorites: There are currently around 16 meteorites (from 20 separate specimens) that almost certainly originate from Mars. They have all crystallised from molten rock, and some appear to have been altered by water on the surface of Mars. The Martian origin is deduced by the age, composition and noble gas components of the meteorites (see p. 58).

TOP LEFT **A polished piece of a pallasite (stony-iron meteorite), about 8 cm across. Pallasites are assumed to be similar to material at the Earth's core-mantle boundary.**

RIGHT **Iron meteorite, 40 cm long by 20 cm wide. Iron meteorites were formed by extensive melting on their parent, and are the closest material to the Earth's core.**

These planets probably started out in the same way as the rocky planets, as an accumulation of rocky material. Temperatures at this distance from the Sun were sufficiently low that water-ice collected along with the dust, while volatile elements were retained and not vaporised. As matter clumped together it formed central cores that were large enough to attract and retain significant amounts of hydrogen and helium from the remaining nebula gas.

Uranus and Neptune, however, contain lower amounts of hydrogen and helium and higher amounts of carbon, nitrogen and oxygen than Jupiter and Saturn. Astronomers think this indicates that Uranus and Neptune formed some time after Jupiter and Saturn, after most of the nebula gas had dissipated.

Beyond the giant gas planets lies Pluto, the smallest planet in the Solar System, with a radius even smaller than that of the Moon. Beyond Pluto is the Kuiper Belt, a region of rocky and icy bodies more like comets than asteroids. (Pluto, indeed, may be a Kuiper Belt object, rather than a planet.) At the outermost reaches of the Solar System, in fact defining the boundary of the Solar System, is the Oort Cloud, the region of space from which comets originate.

Comets

Comets are icy dustballs, or dusty iceballs. They are compact collections of silicate dust and ices containing organic compounds. The icy and volatile-rich nature of comets implies that they have not been heated or melted since they were made during the final stages of the formation of the Solar System. They are, therefore, samples of the most primordial material available for study, and are reservoirs of interstellar and interplanetary dust and ice.

Comets formed at the outer reaches of the solar nebula, way beyond the region where water is a liquid. The central part of a comet, termed the nucleus, is generally only a few

BELOW LEFT **Photo-montage of the four gas giant planets, from the top: Neptune, Uranus, Saturn and Jupiter.**

BELOW RIGHT **Halley's Comet as it appeared in February 1986. The bluer of the two tails (uppermost) is made up of gases, whilst the lower tail, illuminated by reflected sunlight, is the dust tail.**

kilometres across (for example, the nucleus of Halley's Comet is about 16 x 8 km or 10 x 5 miles), while its tail can stretch for hundreds of thousands of kilometres. Most comets never approach the Sun, but remain in stable orbits within the Oort Cloud. Occasionally comets are ejected into the inner part of the Solar System, on orbits around the Sun. When a comet approaches the Sun it develops a dust tail: as the comet's ices vaporise, dust grains that were trapped within them are swept backwards. Some of the dust grains eventually land on Earth, settling through the atmosphere as cosmic dust. Comets played a vital role in the development of life on Earth, and could, therefore, have been instrumental in the

Cosmic dust

The Earth is constantly bombarded by extraterrestrial material, ranging from micron- to metre-sized bodies. Around 50,000 tonnes (49,200 tons) of material fall each year, of which over 90% comes from particles less than 1 mm in size, known as cosmic dust, micrometeorites or interplanetary dust particles (IDPs). Research aircraft operated by the National Aeronautics and Space Administration (NASA) routinely collect cosmic dust from the atmosphere at altitudes of 18–22 km (11–13.5 miles). Cosmic dust may also be collected from the Earth's surface in places that have not been contaminated greatly by terrestrial dust. One of the most successful recovery sites is Antarctica: by melting large volumes of Antarctic ice and then filtering the water, scientists have recovered cosmic dust that has been little altered by terrestrial processes.

Cosmic dust derives from all bodies within the Solar System (including material from planets, their dust rings and satellites), as well as interstellar dust. It forms when asteroids collide and it is ejected from the tails of comets as they journey past the Sun. Cosmic dust is mainly composed of silicates, although much of it is also rich in carbon.

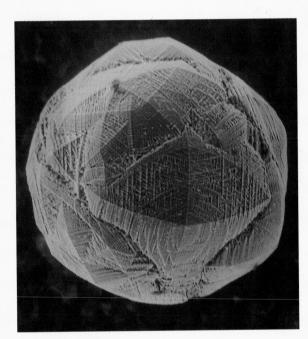

ABOVE **SEM image of a micrometeorite about 70μm in diameter, separated from Antarctic ice. Study of its mineralogy enables us to learn about different types of dust in the Solar System.**

development of life on other bodies in the Solar System. When comets bombarded the early Earth, they deposited their ices and dust on to its lifeless surface. Although the Earth already contained some organic compounds and water, inherited from the primordial dust from which it was built, the Earth received additional organics and water from comets. These were the building blocks of life. Organic compounds from interstellar space also came to the Earth as cosmic dust. The same mechanisms presumably imparted similar materials to Mars, and other potential habitats within our Solar System.

Early history of the Earth-Moon system

The Earth's history, which started with the collapse of part of an interstellar dust cloud into the solar nebula, continued through complex processes as matter clumped together to form the planet. But its development did not stop there: bombardment by asteroids and comets and a constant recycling of crustal materials mean that the rocks now found at the Earth's crust do not represent the original material that formed out of the solar nebula. In order to understand the precursors of the Earth, the only relevant materials available for study in the laboratory are meteorites (see pp. 12–13).

Once the Earth had formed, internal heat from radioactive decay, combined with gravitational energy and collisional energy from asteroid bombardment, kept the planet molten. As the Earth cooled it gradually formed into a metal-rich core and silicate-rich crust-mantle structure (see p. 25).

A significant event in the history of the Earth was the formation of the Moon. Several mechanisms have been proposed for its formation, namely that it formed independently at the same time as the Earth; that it was a rogue asteroid that became captured in the Earth's gravitational field; or that it broke off from the Earth. The currently accepted hypothesis is that in which a Mars-sized body collided with the Earth around 4.51 billion years ago, after the Earth's core had formed. During the giant impact the crust-mantle regions of both the incoming object and the Earth were vaporised and then mixed.

The final turbulent stages of Solar System formation were traced out by intense cratering of the planets by asteroids and comets. Following from the giant impact that formed the Moon, the Earth suffered a prolonged period of bombardment by smaller

ABOVE **The Moon orbiting the Earth; a mosaic of images taken by the Galileo spacecraft as it flew by the Earth in December 1992.**

projectiles. Although no trace of this battering remains on the Earth's surface, having been erased by subsequent geological processes, the scars are visible as craters on the Moon. This epoch of bombardment lasted until around 3.9 billion years ago, and gradually decreased. During this period, the surface of the Earth was hostile to life, as it was heated and melted by the impacts. Over this interval in the Earth's history, an atmosphere may have been built up and destroyed more than once, in cycles of increasing stability punctuated by episodes of bombardment. Gradually, however, the inner Solar System quietened down, the Earth's surface cooled, its atmosphere was retained, oceans formed and conditions were set to allow life to emerge. Of this stage – the interface between the geological and biological history of the Earth – little is known with certainty.

BELOW **Formation of the Moon.**

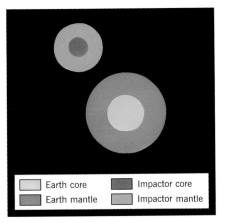

Earth core   Impactor core
Earth mantle   Impactor mantle

A body about the size of Mars comes close to the Earth, after the Earth has formed its core.

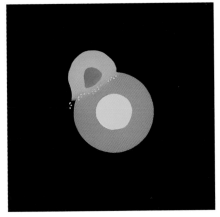

It hits the Earth, vaporising parts of both its own and the Earth's mantle.

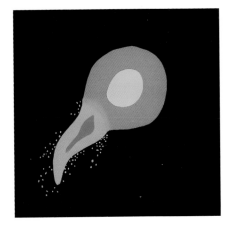

Material is thrown back into space.

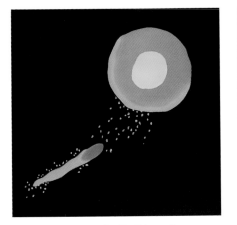

Some falls back to the Earth's surface.

A disk of material is left orbiting the Earth.

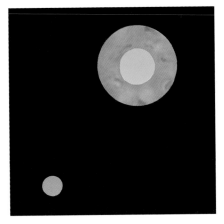

The Moon forms from the disk.

17

What makes Earth special?

So far, all the characteristics that have been discussed apply to many bodies in the Solar System, and not just the Earth: all the planets (and their satellites) have been bombarded by comets and asteroids, carrying their full complement of organic compounds. Seven of the nine planets in the Solar System have moons, and many have, or have had in the past, significant atmospheres. However, the Earth has many special features that set it apart from its planetary siblings within the Solar System; if these features are the reason that life has formed on the Earth, then again

we must look for evidence of the same, or related, properties on other bodies within the Solar System.

Habitable Zone

First of all, Earth is the only planet found in the Habitable Zone (HZ) of the Solar System, the region where temperature and pressure is such that liquid water is stable at the planetary surface. Farther in from the Earth, temperatures are too hot and water evaporates; farther out, temperatures are too cold and water turns to ice. Although the Sun has evolved and become hotter over its 4.56 billion-year

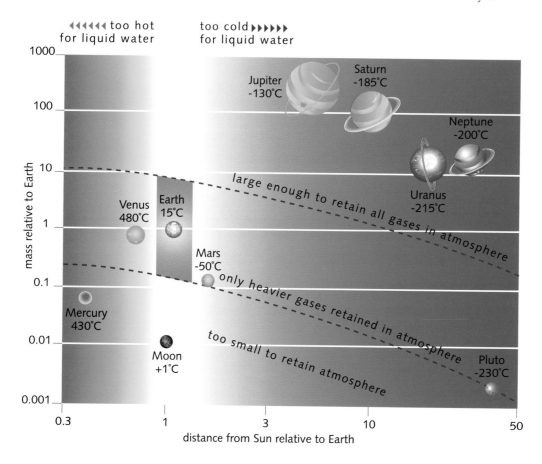

LEFT **Earth is the only planet on which liquid water is stable at the surface. It occupies the Habitable Zone of the Solar System.**

LEFT **Meteor Crater, in Arizona USA, is approximately 1.2 km (0.75 miles) across. It was produced by the impact of an iron meteorite (probably only 40–50 m or about 150 ft in diameter) into the desert about 50,000 years ago.**

lifetime, the Earth has always occupied the HZ. This, however, will not be the case in the future: the Sun is approximately half-way through its 'life', and in around five billion years' time it will have grown into a red giant, with a diameter reaching much farther out into the Solar System than it does currently. By this stage, the surface temperature of the Earth will be too hot to maintain liquid water. Indeed, the HZ might well be in the region of Jupiter or Saturn.

Recently, the concept of the HZ has been amended to take account of factors other than surface temperature and pressure. For example, although the surface temperature and pressure of Mars is too low to permit liquid water at the Martian surface, liquid water might possibly exist within pore spaces between mineral grains in crystalline rocks just below the surface soil layer. Heat sources other than solar radiation also contribute to maintaining a liquid water layer: for instance, on Jupiter's satellite Europa, an ocean of liquid water has been proposed, trapped below a surface layer of ice. The heat source that keeps the water liquid is generated by friction as Europa is alternately pulled and pushed by Jupiter's gravitational field.

Protection from impacts

The Earth is a relatively tranquil place. Although it is occasionally hit by asteroid and comet fragments, the last impact that had a major effect on the global environment was

Jupiter and comet Shoemaker-Levy 9

Jupiter's role as a defence shield was most graphically illustrated in July, 1994, when comet Shoemaker-Levy 9 (D/1993 F2, or S-L 9) entered the inner part of the Solar System. The comet was in 21 pieces, which travelled towards Jupiter in a regular procession (like beads on a necklace, resulting in S-L 9 being dubbed the 'string of pearls' comet), impacting the planet one after another over several days. The impact speed of each fragment was about 60 km per second (135,000 miles per hour). The explosive energy of the largest fragment (fragment G,

the eighth one to hit Jupiter) was approximately 25,000 megatonnes (24,600 megatons) and projected a plume of gas 3000 km (1864 miles) up from the planet. The scar left on the face of Jupiter was about 80% of the diameter of the Earth. If any one of the fragments had hit the Earth it would probably have been sufficient to wipe out humanity, as well as a fair percentage of all other land-dwelling species. And so the Earth has been, and still is, shielded from the worst excesses of asteroidal and cometary bombardment by the protective influence of its giant neighbours.

RIGHT **The scar left behind after fragments of comet Shoemaker-Levy 9 crashed into Jupiter. The shadow is almost the same size as the Earth.**

BELOW **Comet Shoemaker-Levy 9 on its approach towards Jupiter in 1994. The fragments are strung out across a distance of about 1.1 million km (710 thousand miles).**

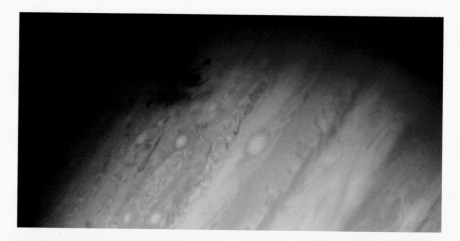

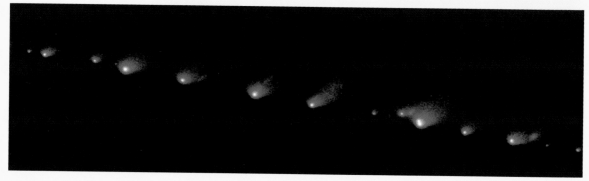

65 million years ago, at end of the Cretaceous period. After that impact, approximately 60% of all known species became extinct, most famously the dinosaurs. Although the Solar System is still a dangerous place, our planet is protected from the worst excesses of bombardment by the presence of the giant planets, whose immense gravitational fields scoop up any incoming bodies: Jupiter, in particular, has been responsible for attracting many such bodies, most recently and spectacularly, the comet Shoemaker-Levy 9.

Stability

The Earth also has a much closer neighbour that exerts a beneficial influence: the Moon. Without our satellite, the Earth would be subject to much wilder swings in climatic variation than is currently the case. This is because the Moon damps down the tendency

ABOVE **The full moon. Darker areas are the 'maria', or regions where rocks have been melted during impacts.**

BELOW **The Earth's axis is tilted towards the Sun.**

of the Earth's axis to vary in response to gravitational tugs from the Sun and Jupiter. The Earth spins completely on its axis once every 24 hours (one 'day') and makes a complete revolution of the Sun every 365.25 days (one 'year'). The axis of the Earth's spin is at an angle of about 23° to the plane of the Earth's orbit (its 'obliquity', see diagram), an effect which gives us our seasons. At point A in the diagram, the Northern Hemisphere is experiencing winter (farther away from the Sun) and at point B the Northern Hemisphere enjoys summer (closer to the Sun). The angle of tilt is fairly stable, and has only varied by a few degrees over the last 600 million years or so. This stability is because the Moon reduces the gravitational pull of both the Sun and Jupiter on the Earth's axis – without the Moon, the axis would be subject to much greater variation, leading to chaotic swings in seasonal climatic change.

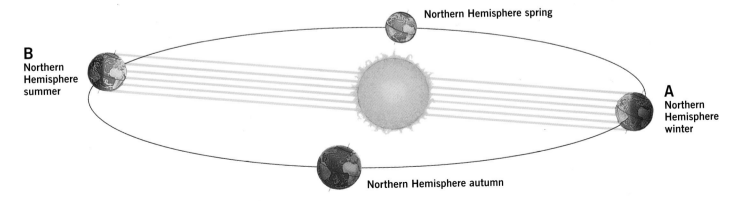

Northern Hemisphere spring

B
Northern Hemisphere summer

A
Northern Hemisphere winter

Northern Hemisphere autumn

Protection from the Sun

As the Earth spins on its axis, and orbits the Sun, it is bathed in solar radiation. The Sun is a stable star: it is burning steadily, giving out energy in the form of heat and light. The radiation level, as measured at the Earth's surface, is just at the appropriate level to maintain the Earth's favourable climate. But without protection, the radiation level would be dangerous: direct exposure to the ultraviolet (uv) component of solar radiation is harmful to most species including human beings, since uv radiation rapidly breaks down bonds between carbon atoms, destroying organic compounds. The Sun is also the source of a second potentially destructive phenomenon known as the solar wind. This is a stream of charged particles that flow outwards from the Sun. The particles can interact with molecules in an atmosphere, stripping the atmosphere away by a process known as 'sputtering'.

The Earth has two lines of defence that protect its surface from excess uv radiation and the solar wind: its atmosphere and its magnetic field. The atmosphere absorbs uv radiation, decreasing the level of harmful radiation that reaches the planet's surface. As well as keeping radiation out, the atmosphere also keeps heat in, maintaining a regular temperature at the surface. Without the atmosphere, surface temperatures would drop to below 0°C (32°F). The Earth's magnetic field, generated by convection currents in the Earth's metallic core, deflects the solar wind. Without its atmosphere and its magnetic field, the surface of the Earth would be inhospitable for the survival of carbon-based life.

A restless planet

The final feature that sets the Earth apart from its planetary siblings is the process of plate tectonics (see p. 25), whereby rafts of rigid rock 'float' on a sea of more mobile rock. Why is this process important for life? Because the process of plate tectonics is part of the global cycle that maintains the carbon balance.

Carbon dioxide is put into the atmosphere by respiration of living organisms, by gases released from volcanoes and hot springs and from breakdown of rocks such as limestones. Carbon dioxide is taken out of the atmosphere by photosynthesis, by dissolution in water and by weathering of silicate rocks. The balance between these competing

RIGHT **The Earth's atmosphere, vital for protecting us from the Sun's radiation, is seen as a thin blue haze in this image taken from the space shuttle.**

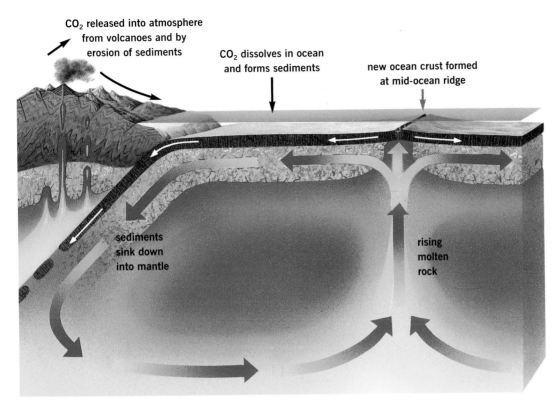

CO₂ released into atmosphere
from volcanoes and by
erosion of sediments

CO₂ dissolves in ocean
and forms sediments

new ocean crust formed
at mid-ocean ridge

LEFT **Cross-section of ocean floor showing molten rock rising beneath the mid-ocean ridge and spreading away either side.**

sediments
sink down
into mantle

rising
molten
rock

processes is extremely sensitive to changes in global temperature, which is directly related to the amount of carbon dioxide in the atmosphere. Without plate tectonics, returning carbon dioxide into the atmosphere at volcanoes, eventually all the carbon dioxide would disappear from the atmosphere, and be fixed into the lithosphere. This would lead to a dramatic drop in global temperature, resulting in the expansion of the Arctic and Antarctic ice-caps to cover much of the ocean and land surfaces. Following on from this, a negative feedback loop would be initiated: increased amounts of sunlight would be reflected back from the Earth's surface (since ice is much more reflective than water or rock), leading to further cooling, formation of more ice, more sunlight reflection, and so on, in an increasing downward spiral to disaster.

The Earth is delicately poised between snowball and greenhouse, and the key to this balance is the fate of carbon dioxide. The balance is maintained by the process of plate tectonics, removing carbon dioxide at subduction zones and producing carbon dioxide at volcanoes and hot springs. As far as we know, the Earth is the only planet in the Solar System to exhibit plate tectonics, although there is some evidence to suggest a limited type of tectonic activity may have occurred on Mars at an early point in its history.

ABOVE **Mountains showing deep gullies and ridges cut by erosion. The uplift of mountain chains and subsequent erosion is an important part of the global carbon cycle.**

From unique Earth to life

The Earth is set apart from all other planets within our Solar System by its unique position relative to the Sun – the only planet with a surface temperature and pressure at which liquid water is stable. The Earth is fortunate that a series of circumstances (its relation to the giant planets, large moon, its atmosphere and magnetic field) have combined to enhance the stability of the planet, allowing an extended period over which life could develop and evolve. Plate tectonics keep the carbon balanced between atmosphere, hydrosphere and lithosphere, so maintaining optimum conditions on the Earth.

All these circumstances conspire to render the Earth and the development of higher forms of life unique, almost certainly in the Solar System, and possibly in the cosmos. Indeed, the 'Rare Earth' hypothesis (proposed by the authors Ward and Brownlee in their book of the same name) has rapidly become something of a paradigm amongst astrobiologists. However, although the Earth is home to an enormous diversity of life, the most ancient, successful and widespread organisms are the bacteria, organisms that presumably had their source in simple molecular building blocks and which might have arisen independently on other bodies in the Solar System. The challenge now is to understand how life originated on the Earth, where it is found and to look for clues to how it might originate and survive elsewhere in the cosmos.

Allied to the significance of plate tectonics in terms of the carbon cycle is the importance of mountain building and the competing process of erosion. At collision boundaries, rocks are uplifted into mountain chains. Gradual erosion of the mountains is countered by continued uplift. If plate tectonics were to cease, then so would this mountain-building activity. Erosion would eventually wear down the mountains. Sediments removed by erosion, carried into the oceans by rivers and streams, would eventually lead to a rise in sea-level. If the process continued the Earth would eventually become covered by a global ocean, resulting in catastrophic extinction of all land-based species.

Plate tectonics

The phenomenon of plate tectonics, or continental drift, derives directly from the internal structure of the Earth. A molten outer metallic core surrounds a solid inner metallic core, overlain by the mantle – a layer of partially molten silicate rocks, above which is the buoyant but rigid lithosphere. The lithosphere is not a continuous layer of material, but is made up of individual pieces, or plates, that fit together like the pieces of a giant jigsaw puzzle. The plates are bodies of slowly moving rock floating on top of the mantle.

The plates are in constant motion, colliding with each other, sliding past each other, or moving apart. Different types of activity occur at the plate boundaries. The region where an oceanic plate collides with another oceanic plate or a continental plate is called a subduction zone. This is where the lithosphere is dragged down into the mantle and melted. Where two continental plates are colliding, rocks are uplifted into mountain chains (for example, the Himalayas, which have formed where the Indian plate moves northwards against the Eurasian plate at a rate of around 5 cm or 2 in per year). Where plates are sliding past each other, transform faults develop, and these are associated with some of the world's major earthquake zones (for example, the San Andreas fault in California, where the Pacific plate is moving eastwards against the north-north-westwards moving North American plate at a rate of about 1 cm or 0.4 in per year). Spreading centres are regions where plates are moving apart and new ocean crust is forming (for example, the mid-Atlantic ridge, where the North American plate is moving away from the Eurasian plate at a rate of about 4 cm or 1.5 in per year). Spreading centres are usually characterised by chains of volcanoes, hydrothermal vents and hot springs.

BELOW **Cross-section of the Earth showing its internal structure.**

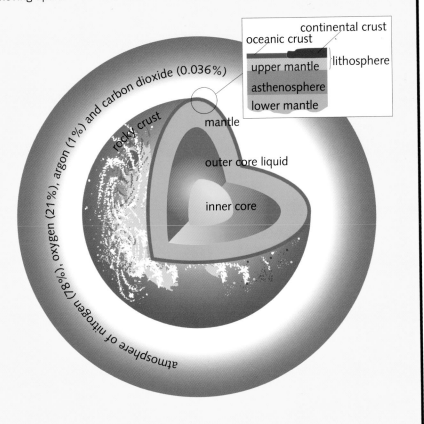

atmosphere of nitrogen (78%), oxygen (21%), argon (1%) and carbon dioxide (0.036%)

rocky crust

mantle

outer core liquid

inner core

continental crust
oceanic crust
lithosphere
upper mantle
asthenosphere
lower mantle

Life on Earth

LEFT **The Sun is a vital source of energy. Water and energy are necessary requirements for life to survive.**

Any attempt to study the origin of extra-terrestrial life has to be based on what we know about the evolution of life on Earth. We assume that the building blocks for life are the same throughout the cosmos, and that the processes operating on these blocks are governed by the same rules of physics and chemistry as on the Earth. There are two key requirements for the emergence of life: the correct ingredients, and favourable conditions.

Recipe for life

You can compare the process of 'building' life to making a cake (see opposite page). It is not sufficient to have the starting materials – they must be mixed together, and supplied with energy before any reaction will occur. Conditions must remain favourable and stable while time passes so that the ingredients can evolve into the finished product, be it a cake, a bacterium or a human being.

What are the 'ingredients', or building blocks for life? On Earth they are the elements carbon, nitrogen, oxygen and hydrogen (four of the five most abundant elements in the cosmos, the fifth being helium), plus phosphorus and sulphur. The most important of these elements is perhaps carbon. Because of its unique atomic structure, carbon forms long chains and rings of atoms to which atoms of the other key elements can attach. The result is an infinite number of different molecules with a variety of properties. All living things on Earth are carbon-based, i.e. composed of carbon chains and rings (known as organic compounds).

No other element has the ability to produce stable chains and rings like carbon. The only element that approaches carbon in its versatility is silicon, an element with a related atomic structure to carbon. Silicon forms extended chains and rings when combined with oxygen, and the silicon-oxygen bond is the basis of the vast inorganic chemistry of the rock-forming minerals.

If, because of its atomic structure, carbon is the basis of life on Earth, then it is only logical to assume that it is also the basis of life elsewhere in the Solar System.

Returning to the analogy of making a cake: the individual ingredients will not, by themselves, turn into cake mixture – they have to be brought together in a mixing bowl. It is the same with the building blocks of life. Simple molecules by themselves will not react to form more complex molecules: they must be brought together by a suitable substrate or solvent.

Recipe for Life

	CAKE	LIFE
Ingredients	Butter, sugar, eggs, flour	Carbon, hydrogen, nitrogen, oxygen, phosphorus, sulphur
Substrate	Mixing bowl	Water
Mechanism	Cream butter with sugar; beat in eggs; add flour gradually	Simple to complex molecules; self-replication; isolation; information transfer
Energy	Bake in a moderate oven	Sunlight; chemical energy
Conditions	Keep oven door closed	Protect from asteroid impact and uv radiation
Time	20 minutes	Bacteria: 700 million years Human beings: 4.5 billion years

Liquid water is the most effective medium for solution and transport of molecules. Although other liquids can also act as solvents for different molecules, no other liquid is stable over such a wide temperature range as water, or exists at a temperature range in which biological processes can also occur. For example, ammonia exists as a liquid at 195–239K (−78°C to −34°C or −44°F to −0.2°F), but these temperatures are much too low to support biological processes. So the presence of liquid water is a prime requirement for the development of life.

So we have the building blocks and a suitable solvent, but there is still something missing – energy. This is needed to drive the reactions that change simple molecules into ever more complex compounds.

The spark of life

Charles Darwin, in a letter to J. D. Hooker in 1871, wrote:

"But if (and oh! what a big if!) we could conceive in some warm little pond, with all sorts of ammonia and phosphoric salts, light, heat, electricity, etc. present, that a protein compound was chemically formed ready to undergo still more complex changes…"

This prescient description encompasses most of the conditions now generally believed necessary for the emergence of life. We do not know with absolute certainty the exact details or mechanisms, but the generally accepted sequence of events that led to life presumably started with simple molecules, possibly brought to the early Earth from space. But before any reactions could occur, there had to be an energy source.

A series of experiments carried out in the 1950s showed that when a mixture of simple molecules, such as carbon monoxide, ammonia and methane, had an electric current passed through them, they formed more complex molecules, including the biologically significant amino acids. These experiments were among the first indicators that organic compounds could be produced from inorganic starting materials by nonbiological processes. How then, does the production of complex from simple molecules relate to processes and conditions that might have existed on the early Earth four billion years ago?

Around this time, the Earth was regularly bombarded by asteroids and comets. These massive impacts released huge amounts of energy into the atmosphere which fuelled reactions between atmospheric gases. When the atmosphere subsequently cooled down, the gaseous molecules dissolved in water droplets that rained down into the oceans. From within this watery environment the molecules could then be built up into more complex organic compounds.

BELOW **In 1953, Stanley Miller and Harold Urey showed that simple molecules could be built into more complex molecules by passing an electric current through them.**

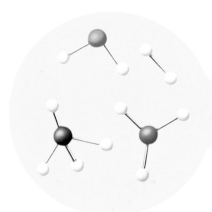

starting materials of simple molecules

electric current passes through molecules

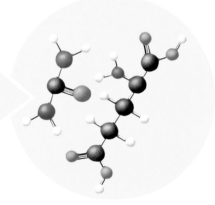

complex molecules created

Where on Earth did life begin?

Owing to the importance of water as a medium for transporting and mixing molecules, we can say that life probably emerged from a watery environment – but whether it was from surface waters or the deep ocean floor we do not know. Until fairly recently, scientists thought that the likely energy source for the reactions that led to the creation of life was the Sun. This implies that organisms originated in surface waters and that the earliest lifeforms relied on photosynthesis – whereby energy from the Sun converts atmospheric carbon dioxide into sugars and carbohydrates, releasing oxygen. The discovery of hydrothermal vents, and their associated lifeforms, on the deep ocean floor has opened up the possibility that the emergence of life might have occurred at depth. It is too dark at the ocean floor for photosynthesis to occur, and so organisms survive by chemosynthesis, whereby energy is derived from chemical reactions. A third location for the first appearance of life has also been proposed: in droplets within clouds. Tiny droplets thrown up from the ocean surface would have been rich in organic materials. When the droplets were carried into the upper atmosphere, reactions between the organic compounds could have been driven by solar energy, producing more complex molecules.

Wherever life began (in surface waters, the ocean floor or in clouds) and whatever type of mechanism drove the process (photosynthesis or chemosynthesis), simple molecules were built up into more complex molecules. The next steps in the evolution of life were the ability to self-replicate, development of an energy conversion process (or metabolism) to drive replication, and a membrane to isolate the molecules from the surrounding medium. It is not clear which of these three items (membrane, metabolism or the ability to self-replicate) came first, or whether they evolved in parallel.

All life today is based on DNA (deoxyribonucleic acid), the double-stranded helix whose structure was first described by

BELOW **Chimneys of sulphide minerals form around vents on the deep ocean floor. The chimneys are home to a rich variety of animals that ultimately depend on the bacteria that also flourish around the vents.**

Watson and Crick in 1953. Many details are still absent from our understanding of how DNA might have come about, but the related molecule, RNA (ribonucleic acid), was probably a significant intermediary in the process (see opposite page). It is hoped that advances in mapping the structure of the humane genome will further the understanding of the origins of life.

What is life?

Before continuing to explore the origins of life, it is useful perhaps to define what is meant by 'life', and how we recognise something as living. The definition of 'life' is not simple, and is an issue that is not only important for understanding how life came about, but also has profound moral, ethical and philosophical dimensions: for instance, when does a molecule, or cell become 'life'?

One dictionary defines life as "the period between birth and death", which, whilst true, is not particularly useful. More apt is the description "life is the sum of all the activities of a plant or an animal". The activities referred to are generally taken to be: respiration, reproduction, nutrition, excretion, locomotion, growth and reaction to external stimuli. Despite the apparent clarity of this definition, it is still not straightforward to apply: there are many systems that follow some of these activities. For example, is a quartz crystal living? Although a quartz crystal may grow, as it builds up layers of atoms, for which 'nutrition' in the form of a feedstock of chemicals is required, and might be thought to reproduce by branching into smaller crystals, a quartz crystal certainly does not

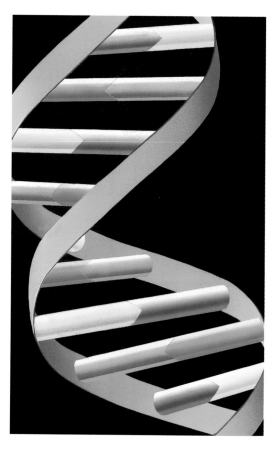

LEFT **DNA, the complex molecule on which all life is based, is a double helix of sugars and phosphates, the two strands of which are held together by a series of smaller molecules.**

move, excrete or react to external stimuli. So it is easy to see that a mineral such as quartz is not alive.

The characteristics of fire, however, have often been taken to illustrate how difficult it is to define something as living. A flame takes in oxygen, and so could be thought to respire. It can grow and move, spread (or reproduce), needs feeding, gives out heat (and so excretes), and reacts to external stimuli, such as a gust of wind. But fire is not 'alive' in the sense taken to apply to organisms. An additional activity, then, is needed for a full description of 'life'. The capacity not just to

RIGHT **Fire shows many of the characteristics associated with living things: it can grow, move, spread and breathe. But fire is not 'living' – it cannot adapt to change, or evolve.**

The RNA world

Deoxyribonucleic acid (DNA) is a repeating pattern of four components (bases) – adenine (A), cytosine (C), guanine (G), and thymine (T) – attached to a long chain of alternating sugar and phosphate groups. The bases are matched in pairs: A with T, and C with G. One suggestion for how DNA, and therefore life, came about from the simple molecules found on early Earth is via the single-stranded intermediary molecule, RNA (ribonucleic acid), which is also made from a backbone of phosphate and sugar groups plus four bases: A, C, G, and uracil (U). When an electric current, such as from lightning, is applied to mixtures of carbon dioxide, carbon monoxide and ammonia (simple molecules found on the early Earth), hydrogen cyanide, cyanoacetylene and formaldehyde are produced. These molecules can be synthesised into the four bases that make up RNA, the precursor of DNA.

The domains of life

It is useful to classify living organisms into groups as it shows links between different species, and the development and sequence of evolutionary patterns. There are many ways of classifying living organisms and for species that are perceived to be higher up the evolutionary ladder, such classification might appear to be straightforward: for instance, it is easy to distinguish plants from animals. However, for a classification scheme to succeed, it must encompass all organisms, not just the most complex. Advances in molecular biology have led in the 1980s to a classification scheme based on the molecular structure of organisms, the Universal Tree of Life, grouping families on the basis of their cell structure. In this system evolution is regarded as following the branches of a tree, rather than going up the rungs of a ladder.

The Universal Tree of Life has three main domains, or super kingdoms: Bacteria, Archaea and Eukarya. Each kingdom has several branches; in the figure, the length of each branch is a reflection of how close it is to its nearest neighbour. Note the proximity of animals to slime moulds!

Although they were originally considered to be the earliest, and so the most primitive of organisms, archaea (or archaeobacteria) are now believed to have adapted through mutation and evolution from even more primitive ancestors, in order to survive in a diverse range of hostile environments. The 'last common ancestor' of all lifeforms appears between the bacteria and the archaea. The first appearance of eukaryotic organisms (with cells that contain a nucleus) occurred two billion years ago, at the same time that Earth's atmosphere became rich in oxygen.

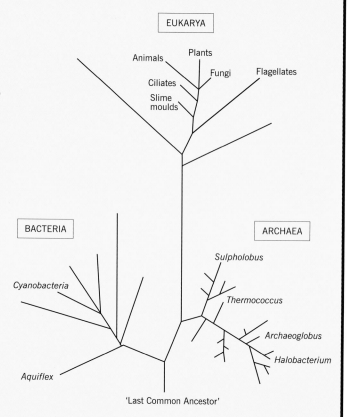

ABOVE **The Universal Tree of Life shows the relationship of different groups to each other.**

grow, but to evolve, is a characteristic of living things, as is the ability to adapt to change, and not just to react to external forces. So we take here the definition that life is the ability to grow, to reproduce, to adapt and to evolve.

When did life begin ?

There are no rocks dating from the original formation of the Earth 4.56 billion years ago on the Earth's surface, as they have been removed through collisions and tectonic processes. But what we do have is a fossil record stretching back to between 3.5 and 3.85 billion years ago. Simple fossils – chains of cells that can be compared to bacteria living today – have been found in 3.5 billion-year-old rocks. Another group of ancient fossils are stromatolites, layered sequences built up by communities of bacteria, that can again be compared with living species today. Going further back in time, to about 3.85 billion years ago, no fossils remain, but rocks in the 3.85 billion-year-old Isua Complex in West Greenland seem to preserve a chemical signature that indicates the presence of biological material.

The Earth was heavily bombarded by comets and asteroids up until about 4 billion years ago, so its surface was too unstable for life to take hold. Since life was apparently in place by 3.85 billion years ago, it seems that

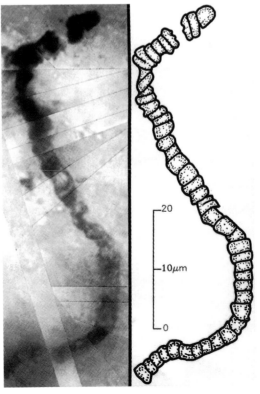

FAR LEFT **Amongst the oldest rocks on Earth is the 3.8 billion years old Amitsoq gneiss from Greenland.**

LEFT **The oldest fossil known on the Earth is the bacterium** *Primaevilium amoenum.*

LEFT **Living algal stromatolites on the seafloor, Exuma Cays, Bahamas.**

LEFT **Fossilised algal stromatolites around 15 million years old, Wadi Kharaza, northern Egypt.**

only around 0.15 billion (150 million) years were required for the successful development of life from simple organic starting materials. However, it took until the start of the Cambrian period (543 million years ago) for the massive expansion in plant and animal species to take place, a period of some 3300 million years. The evolutionary pathway from the earliest simple-celled organisms to primates and humans can now be followed by looking at the structure of the molecules that make up the forms of life.

When the Earth first formed, its atmosphere would have been composed of a mixture of gases added from comets and asteroids as they struck the Earth or emitted from the interior as the planet degassed. The gases were mainly carbon dioxide, carbon monoxide, methane, hydrogen sulphide and steam. As the Earth continued to cool, the oceans formed and the atmosphere became dominated by carbon dioxide. The first fossilised remains are of cyanobacteria, a group of organisms that photosynthesised (took carbon dioxide in from the atmosphere and gave out oxygen). The balance of gases in the atmosphere gradually changed: oxygen built up due to the photosynthesis of bacteria in shallow waters, and carbon dioxide was removed by the formation of limestones, and by weathering of silicate rocks.

The deep ocean was still poor in oxygen, with iron produced in abundance by hydrothermal vents. Ocean currents brought the iron-bearing bottom waters into contact with the oxygen-rich waters of shallower seas; the iron was oxidised and deposited in layered sediments known as banded iron formations (BIF). At this stage much of the oxygen produced by the photosynthesising bacteria was used up by iron oxidation. Eventually, though, oxygen began to accumulate in the atmosphere, and by around 2 billion years ago, deposition of BIF ceased. At this time, there was a sudden increase in the amount of oxygen in the atmosphere, perhaps due to a rise in the amount of photosynthesising cyanobacteria, and a drop in iron productivity at hydrothermal vents.

The limits of life

Once life became established, organisms needed certain conditions to survive, reproduce and evolve. The limits set for the existence of life are based on the physical properties of the components that make up organisms. Ambient temperature, plus the salinity and acidity of water are the most rigorous constraints. Biology is based on bonds between carbon atoms, and also the presence of water as a transport medium. Hence temperature limits

RIGHT **Rocks from the Banded Iron Formation of the Murchison Goldfields in Western Australia. The alternating bands indicate episodes of more or less oxidising ocean waters.**

Physical limits for life on Earth

Parameter	Limiting Conditions	Type of Organism
Water	Liquid water required	
Temperature	–2°C (31.8°F) (minimum)	Psychrophiles
	50–80°C (122–176°F)	Thermophiles
	80–115°C (176–239°F)	Hyperthermophiles
Salinity	15–37.5% NaCl	Halophiles
pH	0.7–4	Acidophiles
	8–12.5	Alkalophiles
Atmospheric pressure	Up to 110 Mpa	Barophiles

OPPOSITE **Strokkur Geyser, south central Iceland, where boiling water and steam are shot high into the atmosphere.**

RIGHT **A hyper—thermophilic archaea which lives in hydrothermal vents at temperatures around 100°C (212°F).**

are defined by, at the lower end, the freezing point of water (dependent on pressure and salinity), and at the upper end by the stability of organic molecules.

In the time between the emergence of life and the present day, microbes have colonised every environment in which it is possible for life to survive. The conditions prevailing in the more extreme 'habitable niches' are taken to define a 'biological envelope' within which life can survive, grow and evolve. These parameters guide the search for life elsewhere within the Solar System. The organisms that survive at these limits are collectively termed 'extremophiles'.

Although there is a diversity of properties exhibited by the micro-organisms, they all have one thing in common: they need water to survive. So what are the most prevalent organisms that live at the extremes, and

where are they found? Many extremophiles belong to the Archaea domain, although there are also significant families of extremophiles in the Bacteria and Eukarya domains.

Thermophiles and hyperthermophiles

Thermophiles are microbes that flourish at temperatures around 50–70°C (122–158°F). Hyperthermophiles are happiest at around 80–110°C (176–230°F); some species cannot grow if temperatures fall below 90°C (194°F). Thermophiles and hyperthermophiles are found all over the world. Hyperthermophilic archaea such as *Sulfolobus*, *Sulfococcus* and *Stygioglobus*, and the thermophile *Thermoplasma* are found on the surface in hot springs and geysers in volcanically active

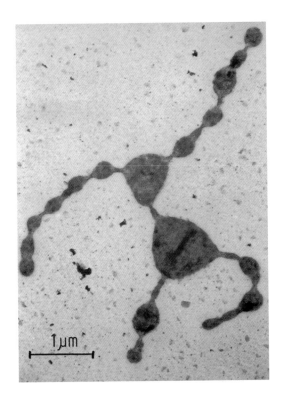

1μm

LEFT **Flamingos feed in the shallow waters at the edge of Lake Nakuru in Kenya. The caustic water teems with alkali-tolerant bacteria that provide a rich food source for the birds.**

regions, fumaroles and solfataras (places where hot gases are emitted from vents, rather than liquids).

On the ocean floor, hyperthermophilic archaea such as *Pyrodictum*, *Archaeoglobus* and *Methanococcus* occur near hydrothermal vents and undersea volcanoes. Thermophiles and hyperthermophiles are also found at depth within sediments and oil reservoirs. Under slightly less hostile conditions, thermophiles have been observed living within compost heaps, landfill sites and spoil heaps.

Psychrophiles

Psychrophiles are cold-loving micro-organisms that flourish at, or around 0°C (32°F). The limit for growth is the freezing point of water, which means that in briny solutions, these micro-organisms can survive at around –2°C (31.8°F).

Psychrophilic bacteria and archaea have been identified at the ocean floor, in the cold bottom waters close to the base of

hydrothermal vents, and also within ice-cores drilled from deep within the Antarctic plateau. In contrast, there are psychrophiles that inhabit the much more desiccating environment of the Dry Valleys of Antarctica, within sandstones and quartzites. These cryptoendoliths (organisms hidden within rocks) are lichens: symbiotic communities of fungi and cyanobacteria or algae. These microbes are of great interest to astro-biologists since they could act as a model for organisms that might survive in similar circumstances within the frozen sub-surface layer of Martian soils.

Acidophiles and alkalophiles

Most acidophiles are also either thermophiles or hyperthermophiles, and are frequently sulphur-loving organisms. The most widespread species are the archaea (such as *Sulfolobus* and *Pyrobaculus*) and the bacteria (such as *Aquifex*). They are found around hydrothermal vents and solfataras, as well as oil reservoirs and spoil heaps. They survive down to pH values of around 0–0.7, and metabolise by turning sulphur into sulphuric acid. The sulphur-rich hot springs and geysers of Jupiter's satellite, Io, would make an intriguing location in which to search for acidophilic bacteria.

In stark contrast to acidophiles, alkalophiles flourish at pH values up to 12.5. The most well-known habitat for alkalophiles is the African soda lake, Lake Nakuru, better known for its amazing flamingo population which feed on the alkalophilic (and halophilic, see below) bacterium *Spirulina*.

Halophiles

Halophiles (for example, *Halococcus*) are adapted to survive in liquids with high salt concentrations (up to around 37.5%). Many halophiles (such as *Natronococcus*, *Natronobacterium*) are also alkalophiles – the alkaline soda Lake Nakuru is also hypersaline, and supports a rich diversity of bacteria and archaea. Within our Solar System, halite (a form of salt) has been found in Martian meteorites, suggesting that brine pools must have been relatively common on the surface of Mars. Rapid desiccation of the surface after loss of its atmosphere might have rendered Mars rich in regions of high salt concentration, in which halophilic micro-organisms might have flourished.

Where on Earth is life now?

Life at the Earth's surface is obvious, and higher organisms have spread to occupy a vast range of environments, from the hot, humid rainforests of the tropics, to the cold,

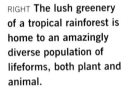

RIGHT **The lush greenery of a tropical rainforest is home to an amazingly diverse population of lifeforms, both plant and animal.**

dry tundra of the polar regions. The range of habitats for micro-organisms, however, is even greater, as less-evolved species colonise niches that are hostile to higher forms of life.

Hydrothermal vents ('black smokers')

A detailed map of the ocean floor shows chains of mountains and ridges. These are spreading centres – areas of intense volcanic activity, regions where molten rock wells up from below and forms new oceanic crust. In the late 1970s, specially adapted submersibles took pictures of the ocean floor for the first time, and found hydrothermal vents (see p. 29). These are elaborate structures, rocky pinnacles, towers and chimneys that seemed to belch out

clouds of black smoke at a tremendous rate. Hydrothermal vents, or 'black smokers', are hot springs: areas where super-heated water (up to 350–400°C or 662–752°F), rich in hydrogen, methane and hydrogen sulphide, shoots up from the sea floor. Where the hot water meets the cold, oxygen-rich bottom water, there is an instant chemical reaction, and sulphides precipitate out from the water, colouring it black. The chimneys build up rapidly from the sulphides, and reach heights of several tens of metres.

Scientists had assumed that the ocean floor would be a barren and desolate region, since no light penetrates the depths, but the discovery of hydrothermal vents showed that

BELOW **Even the cold and desolate tundra of the polar regions supports a host of lifeforms, all specially adapted to drought and low temperatures.**

RIGHT **Although they are screened from sunlight by the ocean depths, the mineral-rich surfaces of hydrothermal vent chimneys are covered with an exotic fauna of tubeworms, amongst which scuttle spider crabs.**

despite the depth and darkness, parts of the ocean floor were home to an unusual collection of animals. Animals such as mussels, crabs and tubeworms colonise the flanks of 'black smokers' feeding on bacteria and archaea. The ocean floor is too dark to allow photosynthesis and so chemical energy is the basis of the foodchain in this strange environment. Micro-organisms such as *Pyrodictum*, *Archaeoglobus* and *Methanococcus* are species of hyperthermophilic archaea found at hydrothermal vents. The micro-organisms flourish at temperatures up to around 113°C (235°F), in neutral or slightly acidic seawater, conditions that occur at the base and around the outside of the vent chimneys. Farther away from the chimney, where temperatures drop to around 3°C (37°F), psychrophiles colonise the ocean floor.

Hyperthermophiles (with very short branches on the tree of life) have only slowly evolved, and might be regarded as still fairly primitive organisms. The recognition of a flourishing ecosystem at hydrothermal vent sites has raised the interesting possibility that life might not have arisen in surface waters, utilising the sun as an energy source, but might have arisen in the deep oceans using chemical energy. A deep ocean origin of life has the advantage that early evolving organisms would be protected from the effects of asteroidal bombardment and solar radiation in the first few turbulent millennia of Earth's history. Discovering communities entirely based on chemosynthesising microbes has also given impetus to the search for life in other deep ocean regions, especially on Jupiter's satellite, Europa, which may have a

LEFT **The wind-scoured Dry Valleys of Antarctica are among the most inhospitable regions on Earth. Despite their low temperatures and desert conditions, they host a remarkable biomass of organisms within the rocks themselves.**

liquid water ocean below a crust of ice. If energy from Jupiter's gravitational pull keeps Europa's core molten and its silicate mantle mobile, then it is possible that hot rocks penetrate to the top of Europa's rocky crust, generating hydrothermal vents that keep the ocean liquid. By analogy with the Earth, Europa's ocean floor might then also be home to its own communities of microbial, or higher species of life based on chemosynthesis.

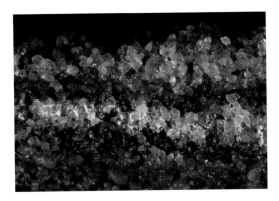

LEFT **A magnified picture of the surface of a sandstone boulder from the Dry Valleys of Antarctica, showing traces of the colony of organisms that survive within the rocks.**

Antarctica

The continent of Antarctica accounts for approximately 10% of the global landmass. Its desert climate (0% humidity) prevents it from supporting a complex ecosystem of plants and animals, although the surrounding ocean waters team with a rich variety of life.

Even so, there is a significant biomass within the sandstones and quartzites that comprise the rocky outcrops of the Dry Valleys of Victoria Land. Cryptoendolithic communities of lichens colonise the top centimetre or so of the sandstones. The communities are layered: the outermost layer of iron-stained quartz is

Panspermia

In the early part of the 20th century, the Swedish chemist Svante Arrhenius proposed the idea of panspermia, where living organisms were spread through interstellar space. By the mid-1920s, this idea had been overtaken by recognition that life might have arisen on Earth from simple organic molecules present either naturally on the early Earth or brought in by comets or asteroids. The panspermia hypothesis has been revived by the astrophysicists Fred Hoyle and Chandra Wickeramasinghe, who have argued that characteristics of interstellar dust measured by telescopes match similar measurements made on fairly complex biological materials, such as cellulose and the bacterium *Escheria coli*. From this, Hoyle and Wickeramasinghe have suggested that the Earth was not just seeded by organic molecules and water from space, but by cells, and even bacteria. Whilst this argument is accepted by very few other scientists, it has served to re-ignite speculation on the viability of organisms in space. Following on from this, recognition that material can come to the Earth from the Moon and Mars as meteorites has raised the question of whether organisms could be transported between planets during asteroidal impact, leading to possible cross-contamination of planets. Could life have evolved on Earth, then got carried out to Mars by impact ejecta, or possibly vice versa? Until we have reliable data on a range of species collected from different planetary environments, it is difficult to address the likelihood or otherwise of the panspermia hypothesis.

barren, but about 1 mm below the surface there is often a narrow, dark band comprised of fungi. The dark pigments block out excessive sunlight and absorb uv radiation. Below the dark layer is a second layer of fungi, this time colourless, under which is a zone of photosynthetic microbes: green or brown algae or cyanobacteria.

Although the Antarctic climate is that of a desert, the environment in which these microscopic communities survive is not particularly dry: water trapped within pore spaces in the sandstones can impart a relative humidity of up to 100%, such that traces of liquid water are present within the rocks. External temperatures are frequently around −30°C (3.8°F), but the ambient temperatures inside the sandstones are warmer, and may be up to 10°C (50°F). The microbial communities are considered to be useful models for the type of biota that might be found within rocks at the Martian surface – a cold, dry, uv-bombarded windy desert not too different from the Dry Valleys of Antarctica.

There is another group of micro-organisms living in the hostile desert of Antarctica, this time in the ice. The sub-surface freshwater Lake Vostok is one of almost 80 lakes that have been identified by radar mapping of the Antarctic plateau. The lakes are buried below layers of ice up to 4 km (2.5 miles) in depth. Ice cores drilled into the ice have recovered viable micro-organisms (bacteria, fungi and algae) at depths down to 2.5 km (1.5 miles). The ice at this depth is around 400,000 years old, and presumably has always been sealed from the outside environment. Astrobiologists

believe that Lake Vostok may be a useful terrestrial model for the ice-covered Jovian satellite Europa; the survival of bacteria deep within Antarctic ice implies that micro-organisms might also be able to survive in a similar environment on Europa.

Hot springs and solfataras

Hot (sometimes boiling) water, acidity and high sulphur concentrations are characteristic of the hot springs and geysers which frequently mark regions of volcanic activity across the Earth's surface. Water temperatures are generally 40–90°C (104–194°F). Despite the temperature and acidity, a rich ecosystem survives at the base of the geyser. Solfataras are also associated with regions of volcanic activity, but steam and hot gases (especially sulphur dioxide) are emitted, rather than boiling water; conditions are still highly acidic.

From life on Earth to life elsewhere

Although we can sketch a framework for how life originated on the Earth from simple molecules, via complex molecules, to the formation of RNA and eventually DNA, details of the precise chemistry, mechanism, timing and location of this momentous event (or series of events) are still unknown. Using the genetic make-up of organisms, we can look back through the 'Tree of Life' to find, if not its rootstock, at least its lowermost branches. The most ancient organisms are the bacteria, which took hold on Earth very early on in the planet's history, only about 700 million years after the Earth formed. Over the

OPPOSITE **A hot spring at Waiotapu, New Zealand. A variety of organisms live around the margins of the spring, despite, the high temperature of the water.**

following 3.85 billion years, micro-organisms have colonised almost every niche available to them, from the driest deserts of Antarctica, to the deep ocean floor, and the scalding, sulphurous and acidic waters of hot springs and geysers. Given the range of habitats occupied on the Earth by these primitive organisms, then the search for equivalent niches on other bodies, both within and beyond the Solar System, might be similarly rewarded by the discovery of a whole host of extraterrestrial lifeforms. So, armed with a knowledge of where micro-organisms

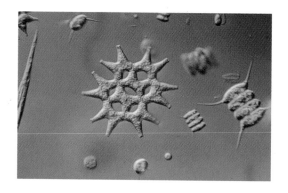

LEFT **Just a single drop of pond water teems with an abundance of organisms.**

flourish on Earth, we can start to pinpoint likely looking locations on our neighbouring planets and their satellites.

Life elsewhere within the Solar System

A voyage through the Solar System to search for traces of past or present life is, in effect, a search for liquid water. The concept of a Habitable Zone (HZ) around a star was originally thought of as a fairly narrow region where the surface temperature of a planet was appropriate for liquid water to be stable. Within our Solar System, only the Earth orbits at the appropriate distance from the Sun for this condition to be met. This is the 'Goldilocks hypothesis', where conditions are not too hot or too cold, but are just right. Even though recent models predict a slightly wider HZ around the Sun and similar stars, the HZ does not reach out as far as Mars or Jupiter. But there are other effects to take into account when considering the possibility of viable habitats on these planets and their satellites, such as the presence of thicker atmospheres, or heating due to the gravitational pull of a planet (tidal heating).

Why should life be elsewhere?

But why should water, or indeed any of the ingredients necessary for life, be present on other bodies? The answer lies far back in time, during the final stages of the formation of the Solar System. Around four billion years ago, the Earth suffered intense bombardment by asteroids and comets. Although the Earth's surface would initially be rendered sterile by the bombardment, it is possible that incoming bodies delivered water and organic compounds to the surface of the Earth,

allowing the atmosphere to stabilise, oceans to form and providing the ingredients for life. As the impacts decreased in frequency, life was able to take hold.

As we have discussed, comets are icy bodies, rich in organic compounds – if they could deliver such materials to the Earth, then

ABOVE **Photo-montage of eight of the planets in the Solar System, plus the Earth's Moon. Missing is the smallest, and farthest planet, Pluto, which has not yet been visited by spacecraft.**

47

To Boldly Go: the naming of space missions

The first craft to be launched successfully into space was the Sputnik satellite, by the (former) Soviet Union, on 4th October, 1957. 'Sputnik' is often translated as 'satellite', but is perhaps more appropriately translated as 'companion', or 'accompaniment', reflecting the close relationship between the artificial satellite and the Earth. From Sputnik onwards, a great deal of attention has been paid to the naming of space missions.

Often the meaning behind the names is self-evident: Ranger, Surveyor, Pioneer, Mariner, Voyager and Pathfinder, for instance, were all spacecraft with a prime mission objective of exploration. Stardust (launched in 1999) is a mission to collect dust from stars and a comet, and Luna was the Russian Moon exploration programme. Sometimes an acronym suffices, when it spells out a recognisable word, such as FIRST, NEAR, SOHO, although it does not necessarily inform non-specialists about the purpose of the mission. FIRST is a Far-InfraRed Space Telescope, and NEAR is the Near Earth Asteroid Rendezvous mission, while SOHO is a shortened form of the Solar Heliospheric Observatory.

Sometimes the reasoning behind the selection is obscure: why was Apollo (Greek and Roman god of the Sun) chosen as the name for the lunar exploration programme? A search of NASA's archives is unforthcoming on this point. Mars Express is the European Space Agency's rapid mission to Mars in 2003, but why is its lander called Beagle 2? Charles Darwin sailed on the Beagle when he made the observations that allowed him to write On the Origin of Species, in which he first proposed the theory of evolution. It is hoped that when Beagle 2 lands on the surface of Mars in late 2003, the instruments on board will collect data that will shed light on the possibilities of life on Mars. Giotto was ESA's first mission to a comet, flying close to Halley's Comet in 1985–86. One of the most famous illustrations of Halley's Comet is by the artist Giotto, where the comet is depicted as the Star of Bethlehem in his painting of the Nativity. Rosetta is ESA's second cometary mission, designed to unlock the secrets of primitive Solar System material, in an analogous fashion to that in which the Rosetta Stone gave archaeologists the key to translating Egyptian hieroglyphics, leading to a greater understanding of ancient civilisations.

Yet a third group of names relates to famous astronomers who have had some connection with the target object of the mission. So Galileo Galilei (the 17th-century Italian astronomer who described Jupiter's four largest satellites) is honoured by the Galileo mission that is mapping those same satellites at high resolution. The Hubble Space Telescope (HST) is named after Edwin Hubble, the American astronomer who first recognised that many galaxies existed in the Universe, not just one, and that these galaxies were all moving apart. Cassini and Huygens, two great contemporaries of 17th-century astronomy, are jointly recognised in the combined ESA-NASA mission to Saturn and its giant satellite Titan. Cassini described Saturn's rings and discovered four of its satellites, whilst Huygens was the discoverer of Titan.

Adventurers throughout history have also been commemorated: Ulysses, the great explorer of Greek legend, is a mission charting unknown regions of space. Magellan, the 16th-century Portuguese explorer who first circumnavigated the Earth, was the name given to NASA's high-resolution mapping expedition that circumnavigated Venus. The seafaring Vikings, who charted vast parts of the northern hemisphere between AD 800 and 1100 are remembered in one of NASA's most successful planetary missions, the Viking missions to Mars in the mid-1970s.

And boldly going where no telescope has gone before will be the Next Generation Space Telescope, the follow-up to one of the greatest star-trekkers of all, the HST, and named, of course, after the successor of the TV series *Star Trek*.

BELOW **Lift-off of the Apollo 11 space vehicle from the Kennedy Space Center on July 16th 1969, the first mission to land people on the Moon.**

they could also deliver them to the other planets and their satellites. Therefore, there is the potential for life to have started in other habitable niches throughout the Solar System. Whilst Mercury, Venus and the Moon are almost certainly barren of life now, each has features that imply that at times in the history of the Solar System, conditions might have been amenable enough to allow life to arise. Mars has long been favoured as a possible host for extraterrestrial life; more recently, the Galilean satellites of Jupiter (particularly Europa) have been proposed as good places to search for traces of life, while Titan, the giant satellite of Saturn, exhibits many properties that relate to conditions at the surface of the early Earth before life had emerged. Each of these bodies will be considered in turn, in order of their distance from the Sun. We will examine their potential for harbouring life, or at least the seeds of life, if not now, then in the past.

Mercury

Mercury is a rocky planet, with a density similar to that of the Earth. It is, however, much smaller than the Earth, implying a much higher overall metal content; Mercury's core comprises approximately half the total volume of the planet. The proximity of Mercury to the Sun results in daytime temperatures of around 450°C (842°F), but its almost total lack of an insulating atmosphere means that night-time temperatures drop down to –180°C (–146°F). The planet is also engulfed by fierce solar radiation. The hostile environment has usually been taken to suggest that it is most

unlikely that Mercury is home to any type of organism. Interest in Mercury has recently been revived, however, following radar measurements of the polar regions, where brightly reflecting areas were identified as ice-rich. Water-ice is thought to be present in deeply shadowed craters that are never warmed by the Sun.

Venus

LEFT **A mosaic of radar images of Venus, centred on the north pole, taken by the Magellan spacecraft in its 1990–1994 mission.**

Venus is twice as far away from the Sun as Mercury and it, too, has traditionally been regarded as completely hostile to any form of life. Venus is approximately the same size and density as the Earth, but has the hottest surface temperature in the Solar System, at around 480°C (896°F). The high temperature is a result of the planet's very dry and dense atmosphere of carbon dioxide, around 90 times as thick as that of the Earth, with a significant content of sulphur (as sulphur dioxide and sulphuric acid). The greenhouse effect of the insulating atmosphere, which allows uv radiation in, but prevents heat from escaping, has left Venus with a surface temperature even hotter than Mercury. The reason for the greenhouse conditions, and the planet's failure as the potential host for life, is its lack of plate tectonics. As discussed in the second Chapter, plate tectonics on Earth controls the amount of carbon dioxide in the atmosphere. Venus has no plate motion, and therefore no mechanism by which carbon dioxide can be removed from its atmosphere. Gradually, over time, the amount of carbon dioxide has built up, eventually precipitating the runaway greenhouse conditions that exist today. But, far back in Solar System history, the surface conditions on Venus must have been very different. Four billion years ago, the Sun was cooler than it is now, and the planet's atmosphere had a higher water vapour content. Whilst it is unlikely that even at that time Venus was within the solar Habitable Zone, its surface temperature would have been much lower, perhaps even low enough for life to have taken hold. Venus today, therefore, represents an extreme example of the effects of global warming.

The Moon

The Moon is the only body in the Solar System that has been explored first hand by astronauts, and which has had material returned directly from it to the Earth. The Moon has been closely studied by generations of astronomers, and served as an inspiration for science fiction writers and film directors searching for a handy place from which to dispatch (usually unfriendly) aliens. We know much about the Moon's geology, history,

chemistry and atmosphere, and have concluded that it is a dry, desolate and barren body. Although lunar meteorites and the samples from the Apollo and Luna missions showed no signs of alteration by water from processes on the Moon's surface, the two most recent lunar missions, NASA's Clementine (1994) and Lunar Prospector (1998), both detected the apparent presence of water-ice deep within craters at the Moon's poles. At the poles part of the crater walls are permanently shadowed, and it has been surmised that volatiles from cometary and asteroidal bombardment have accumulated in such regions. While there is no suggestion that life might be associated with these discoveries, the calculation that up to 6 billion tonnes (5.9 billion tons) of water might be frozen into lunar soil at the poles has revived speculation of the use of the Moon as a base for future colonisation.

Mars

Ever since the *canali* of Schiaparelli's 19th-century map of Mars were mistranslated

BELOW **The surface of Mars, as sketched by Schiaparelli in 1877, showing the 'canali' he observed.**

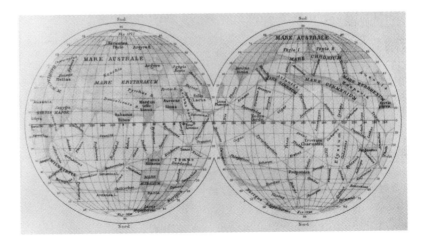

into English as canals (rather than grooves or channels), there has been speculation that life might have existed on Mars. Following on from the first mission, the failed launch of the Soviet spacecraft Marsnik 1 in 1960, over 30 missions have attempted to explore the Red Planet – with varying degrees of success. Although the canali were subsequently found to be mainly illusions resulting from the poor optics used to observe Mars, later images returned by spacecraft clearly showed that the Martian surface was marked by features similar to those carved out by rivers and streams on the Earth's surface.

Like the Earth, Mars is a rocky planet. It has a radius approximately half, and a mass around one tenth that of the Earth; consequently, gravity on Mars is only about 40% that of the Earth. The atmosphere on Mars is also different: it is much thinner than Earth's and is predominantly carbon dioxide (about 95%) rather than nitrogen. The thin atmosphere provides the Martian surface with little protection from heat loss – the average daily temperature is around –60°C (–26°F). Temperatures may reach +30°C (86°F) at the equator in summer, and fall to –130°C (–96°F) at the poles in winter. The structure of Mars is similar to the Earth (with a core and mantle), but it seems to have a rigid crust rather than the more flexible plate structure of the Earth, although recent results from the magnetometer on NASA's Mars Global Surveyor indicate that in the past there may have been limited plate movement on Mars.

Mars exhibits extreme examples of the features shown by the Earth. For example, the biggest volcano in the Solar System occurs on Mars; Olympus Mons is a shield volcano almost three times as high as Mount Everest. Shield volcanoes on Earth (e.g. Mauna Lon, Hawaii) are formed as piles of magma that build up when the crust is above a hot spot in the mantle. But plate movement on Earth prevents continuous accumulation of a magma pile in a single location. On Mars, however, where there has been little or no plate movement, the volcanoes simply increase in size.

The surface of Mars

The first detailed maps of the surface of Mars were produced with data from NASA's

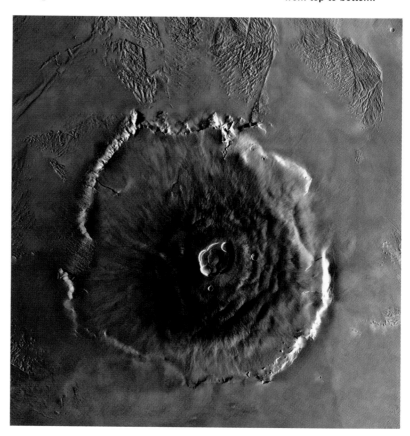

BELOW **Olympus Mons, the largest volcano in the Solar System, measures about 600 km (375 miles) from top to bottom.**

Missions to Mars

Spacecraft	Encounter Date	Mission	Agency
Mars 1960A	*Launch failure, Oct 1960*	*Flyby*	*Soviet*
Mars 1960B	*Launch failure, Oct 1960*	*Flyby*	*Soviet*
Sputnik 29	*Failed, Oct 1962*	*Flyby*	*Soviet*
Sputnik 31	*Failed, Nov 1962*	*Flyby*	*Soviet*
Mars 1	*Failed, June 1963*	*Flyby*	*Soviet*
Mariner 3	*Failed, Nov 1964*	*Flyby*	*NASA, US*
Mariner 4	July 1965	Flyby	NASA, US
Zond 2	*Failed, Apr 1965*	*Flyby*	*Soviet*
Zond 3	July 1965	Flyby	Soviet
Mariner 6	July 1969	Flyby	NASA, US
Mariner 7	Aug 1969	Flyby	NASA, US
Mars 1969A	*Launch failure, Mar 1969*	*Orbiter*	*Soviet*
Mars 1969B	*Launch failure, Apr 1969*	*Orbiter*	*Soviet*
Mariner 8	*Launch failure, May 1971*	*Flyby*	*NASA, US*
Kosmos 419	*Failed, May 1971*	*Orbiter and Lander*	*Soviet*
Mariner 9	Nov 1971	Orbiter	NASA, US
Mars 2	Dec 1971	Orbiter (failed lander)	Soviet
Mars 3	Dec 1971	Orbiter and Lander	Soviet
Mars 4	Feb 1974	Flyby (failed orbiter)	Soviet
Mars 5	Feb 1974	Orbiter	Soviet
Mars 6	Mar 1974	Lander	Soviet
Mars 7	Mar 1974	Flyby (failed lander)	Soviet
Viking 1	July 1976	Orbiter and Lander	NASA, US
Viking 2	Sept 1976	Orbiter and Lander	NASA, US
Phobos 1	*Failed, Jan 1989*	*Orbiter; Phobos Lander*	*Soviet*
Phobos 2	Jan 1989	Orbiter (Phobos Lander failed)	Soviet
Mars Observer	*Failed, Aug 1993*	*Orbiter*	*NASA, US*
Mars 96	*Failed, Nov 1996*	*Orbiter and Lander*	*Soviet*
Pathfinder	July 1997	Lander	NASA, US
Global Surveyor	Nov 1997	Orbiter	NASA, US
Climate Orbiter	*Failed, Sept 1999*	*Orbiter*	*NASA, US*
Polar Lander; Deep Space 2	*Failed, Dec 1999*	*Lander; Penetrators*	*NASA, US*
Nozomi (Planet-B)	Launched July 1998; arrives Jan 2004	Orbiter	Japan

ABOVE **The surface of Mars at the Pathfinder's landing site, showing boulders of different colours and shapes.**

In 1976, NASA's two Viking landers sent back many images of the planet's landscape, showing panoramic scenes of broken boulders distributed over flat, dusty plains. Viking also measured the elemental composition of the atmosphere and surface soils. In combination with the Mariner data, the Viking results allowed scientists to build up a picture of Mars as a rocky planet with a significant geological history. The two hemispheres of Mars exhibit different geological histories. Most of the northern hemisphere consists of almost flat, low-lying plains, showing little cratering. In contrast, the terrain of the southern hemisphere appears more ancient, with cratered highland regions cross-cut by canyons, channels and valley networks.

But many questions about Mars remained unanswered, in particular whether it ever had river systems. The Pathfinder mission of 1997 landed on a rocky plain at the mouth of the Ares Vallis in Chryse Planitia. It recorded spectacular images of a rock-strewn plain, with tantalising glimpses of rounded pebbles and what could be layered structures and hollows within some of the rocks. Sojourner rover, a mobile robotic probe, collected chemical and image data for rocks and soil which implied that some of the rocks might be sedimentary. Paler, almost cream-coloured patches in the soil might be areas leached by fluid action, or where water has evaporated leaving behind salts. The poorly sorted landscape of part-rounded pebbles and boulders could also be interpreted as the type of landscape remaining after catastrophic flooding, further evidence of water at some time in Mars's past.

Mariner 9 orbiter mission of 1971–72, in which channel and valley networks, volcanoes, canyons and craters were photographed at reasonable resolution for the first time. Mariner 9 also returned pictures of layered terrain in Mars's polar regions – which could indicate rock that is sedimentary (i.e. laid down in water). The channel and valley networks bear such striking similarities to river valleys and deltas on the Earth that they were assumed to demonstrate beyond doubt that at some time in its history, Mars must have had significant quantities of running water coursing across its surface.

The most recent images of Mars's surface
have come from the Mars Orbital Camera
(MOC), on board the Mars Global Surveyor.
Many features imaged by Mariner 9 and
Viking have been photographed by the MOC
at much higher resolution (or magnification),
showing the channels and valley networks in
greater detail. What has emerged from these
images has been a re-interpretation of these
surface features. When satellite images are
taken of river valleys on the Earth, as the
resolution of the images is increased an
increasingly fine structure is observed, with

ABOVE **Networks of fine
run-off channels on
steep slopes of ancient
Martian terrain, as
photographed by
Viking 2 in 1976.**

LEFT **Channels cutting
the wall of the Hellas
impact basin on Mars,
photographed by the
Viking 1 orbiter in 1976.**

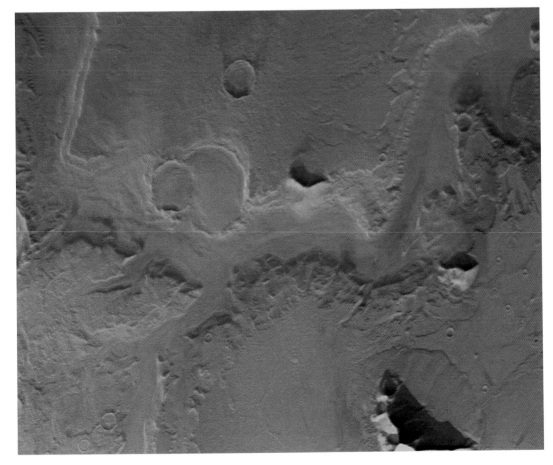

tributaries joining the main flow of the river. The same effect is not seen on Mars: higher resolution images of the channels show that very few 'tributaries' flow into the main pathways. The images have been interpreted as indicating that either the channels were formed by ice, rather than liquid water, or that the flow was below the surface of the planet. Whichever interpretation is correct, the implication is still that water was relatively abundant in surface or sub-surface locations at some time. Exactly when is also a matter for debate. Again, images from the MOC suggest that liquid water might have percolated to the Martian surface in much more recent times than was previously thought, perhaps as recently as one million years ago, and indeed might even still be present in sub-surface locations today.

Interpretation

For water to have been present at Mars's surface, the Martian atmosphere must have been much thicker, and surface temperatures much warmer than they are today. A thicker atmosphere provides greater protection from solar radiation; in a previous epoch, Mars would have had a warmer and wetter climate and provided all the conditions suitable for the emergence of life. Given this framework, it is not surprising that so much interest has been focused on a potential Martian biosphere.

Images showing that the surface environment of Mars has been more amenable for life in the past inspired experiments on the Viking landers. Both craft carried a gas chromatograph-mass spectrometer (GC-MS) and three separate biological experiments.

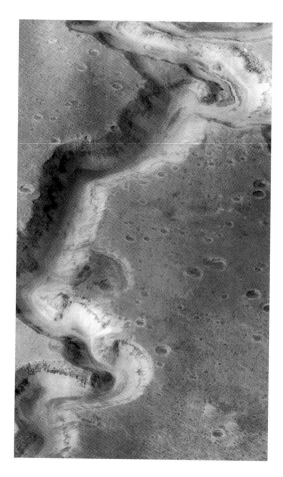

LEFT **The Nanedi Vallis, a channel carved by flowing water or ice cutting through cratered terrain in Xanthe Terra, as photographed by the Mars Global Surveyor Mission in 1999. The channel is 2.5 km (1.6 miles) wide.**

The GC-MS was designed to test surface soils for the presence of organic compounds. The biological experiments were designed to test for metabolic action, which would indicate the presence of life. Unfortunately, the results obtained were, on balance, negative. Although one of the experiments gave a positive signal that might have implied the presence of a metabolising agent, the overall conclusion from the mission was that there was no detectable trace of organic matter in the surface soils at either of the two landing sites, and thus no likelihood of the presence of life.

Future exploration of Mars

Despite these conclusions, Mars remains a prime target in the search for extraterrestrial life. The discovery of micro-organisms on Earth that can survive (and thrive) in conditions previously thought hostile to life, such as within rocks in the Dry Valleys of Antarctica, has given impetus to remote Martian exploration. Since the Viking mission, there have been two successful missions to Mars (NASA's Pathfinder and Global Surveyor). Although neither of these missions included experiments specifically intended to test for traces of past or present life at the Martian surface, they are part of NASA's programme of missions leading up to the return of samples from Mars to Earth.

However, before the sample return mission takes place, there are several other missions scheduled to visit Mars. The Japanese Space Agency have a spacecraft already on its way: the orbiter Nozomi (meaning Hope) will arrive in January 2004. Just prior to that, the European Space Agency's Mars Express mission, to be launched in June 2003, will land on Mars in late December 2003. Its lander, Beagle 2, has a primary aim to search for traces of life in Mars's surface and sub-surface sediments. NASA's Mars 2003 mission consists of two craft, to be launched in April and May 2003, landing on Mars in late spring 2004. On board each lander will be a vehicle, the Athena Rover (pictured), capable of travelling for several hundreds of metres

LEFT **NASA's Athena Rover. Two of these vehicles will land in different regions of Mars in 2004, carrying cameras and other instruments for examining the Martian surface.**

away from the landing platform, and carrying an array of scientific instrumentation to analyse the Martian surface.

Meteorites from Mars

Information obtained from Martian meteorites has been vital to the interpretation of much of the chemical data returned by Viking and Pathfinder. There are (currently) 16 Martian meteorites, 20 specimens in total, known as SNCs (named after three sub groups, Shergotty, Nakhla and Chassigny). They are distinguished from 'regular' meteorites (that originate in the Asteroid Belt) on the basis of their younger (and therefore planetary) ages down to 165 million years. Evidence that the SNCs originate from Mars comes from gases trapped within the meteorites. The Viking landers of 1976 returned data on the chemical and isotopic composition of Mars's atmosphere. Some of the SNCs, such as EET A79001, contain pockets of black glass. The glass was formed by shock melting of mineral grains during the impact event that broke the meteorite from its parent. Measurement of gas trapped within the glass shows that it is identical in composition to that of the atmosphere of Mars. And the only way that Mars's atmosphere could have been introduced into the glass is if the rocks came from Mars.

Scientists anticipated that when Sojourner measured the composition of rocks on the Martian surface, it would find them to be similar to the SNCs (that is, rich in magnesium- and iron-bearing minerals). One of the biggest surprises to come from the Pathfinder data was that the rocks at the landing site were much more silica- and calcium-rich than the SNCs, implying that the Pathfinder rocks had been through several cycles of heating and melting, and had not simply crystallised from a single magma. So there is a diversity of rock-types on Mars.

Unfortunately, Pathfinder returned only chemical and not mineralogical data, so the exact rock types encountered by Sojourner cannot be identified with certainty. Part of the problem arises because only the exterior, weathered surfaces of the boulders were examined. Geologists are familiar with the requirement to break open a rock to look at a fresh surface – this, of course, was not possible on the Pathfinder mission.

RIGHT **Iron- and magnesium-bearing minerals inside the Nakhla Martian meteorite. Image about 2 mm across. The crystal structure indicates crystallisation close to the surface of its parent-body.**

FAR RIGHT **The sawn surface of the EETA 79001 Martian meteorite. The dark patches are shock produced glass that contains trapped Martian atmosphere.**

RIGHT **Diagram showing that the gases trapped inside EETA 79001 are the same as, and in the same relative proportions as, those in Mars's atmosphere.**

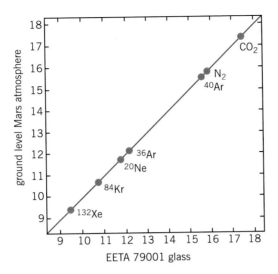

RIGHT **Diagram showing that the gases trapped inside EETA 79001 are the same as, and in the same relative proportions as, those in Mars's atmosphere.**

Until a future space mission brings rocks back to Earth from the surface of Mars, the only material from Mars that is available for direct investigation are the SNC meteorites. Study of minerals within Martian meteorites has revealed how much Mars's atmosphere has changed since the planet formed, and how much water has flowed across the planet's surface.

Evidence of life?

The observation that at least one of the SNCs, EETA79001, contained Martian organic material associated with carbonate minerals sparked a debate on whether this was evidence for extraterrestrial life. This work was followed, in 1994, by the discovery of higher levels of organic carbon in a second Martian meteorite found in Antarctica, Allan Hills (ALH) 84001. ALH 84001 became a further focus of interest in 1996 when a team of scientists, led by David McKay of NASA's Johnson Space Centre in Houston, discovered microscopic features within the abundant

carbonate patches in ALH 84001, and claimed to have found evidence for a primitive 'fossilised Martian biota' in the meteorite (see p. 60). However, identification of the microfossils remains controversial, since much of the evidence is circumstantial and relies on the coincidence between a number of otherwise unrelated characteristics of the meteorite. Even so, the abundant carbonates within ALH 84001 almost certainly formed on Mars, most probably on a part of Mars's surface where water was slowly evaporating from small pools, an environment that is compatible with one in which micro-organisms might survive.

Research shows then that the Red Planet has had ice and liquid water flowing across or just under its surface during past epochs. Components isolated from Martian meteorites imply that when water was present on the surface, it was warm and briny, and restricted in flow. In other words, it might have been locked in enclosed basins that occasionally overflowed in episodes of flash flooding. Although the surface of Mars is now apparently dry, and, as a result of oxidation by solar uv radiation, presumably devoid of organic compounds, we have no knowledge of what is present in the subsurface soil layers. There might be reservoirs of permafrost, or even liquid water trapped within pore spaces, giving rise to organisms like the cryptoendolithic bacteria of the Dry Valleys of Antarctica (see p. 43–44). Residues from evaporating brines might host desiccated sulphur-loving micro-organisms. But in the absence of suitable meteoritic material, the answer to the question of the existence (either

Are there microfossils in Allan Hills (ALH) 84001?

The ALH 84001 Martian meteorite contains several different carbon-bearing compounds including organics and carbonates. The organic molecules are a mixture of material from Mars contaminated with organic molecules from the Earth. The carbonates occur as bright orange-coloured rounded patches, and were formed on Mars from water that had carbon dioxide from Mars's atmosphere dissolved in it. A team from NASA, led by David McKay, also studied ALH 84001 and made several observations that caused them to believe that ALH 84001 contained fossilised martian lifeforms. These observations were: (1) part of the organics were a particular type of compound that, on Earth, is produced when living material fossilises; (2) tiny grains of magnetite (iron oxide) were present, with shapes unlike those of magnetite grains in meteorites, but similar to magnetite produced by living organisms; (3) shapes on the surface of the carbonate patches resembled

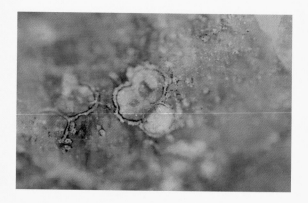

bacteria. In summarising their findings, McKay and his team were careful to point out that each observation was capable of an alternative explanation involving an inorganic process, but that the most likely explanation that satisfied all the observations was that the features were 'the fossil remains of a past Martian biota'. This opinion is accepted by very few researchers, but study of the features in ALH 84001 continues, and has stimulated investigation of the possibilities of life on Mars.

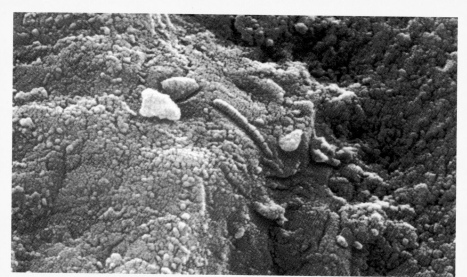

ABOVE **Patches of orange carbonate in ALH 84001. Size of image about 3 mm.**

LEFT **SEM image (size about 1μm) of the surface of the orange carbonates in ALH 84001, showing a 'microfossil'.**

now or in the past) of life on Mars must await results from future space missions to the Red Planet.

The Jovian satellites

Beyond Mars, and beyond the Asteroid Belt, lies Jupiter, the largest planet in the Solar System. Jupiter is a gas giant with a radius approximately 11 times and a mass around 320 times that of the Earth. Jupiter's composition is close to that of the Sun, mainly hydrogen, with about 14% helium plus minor amounts of the gases methane, water, ammonia, ethane, ethylene, carbon monoxide, hydrogen sulphide and phosphine. The turbulent nature of Jupiter's atmosphere is seen by movement in the cloud belts that circle the planet, and the huge bolts of lightning that have been recorded by spacecraft flying close to the planet.

LEFT **Global view of Jupiter, showing its Great Red Spot.**

The Galilean satellites of Jupiter

Cross-section through the Galilean satellites, showing their inferred internal structures.

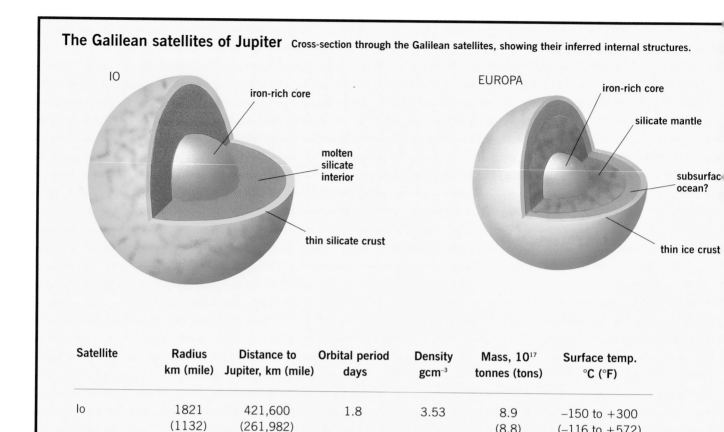

Satellite	Radius km (mile)	Distance to Jupiter, km (mile)	Orbital period days	Density gcm⁻³	Mass, 10¹⁷ tonnes (tons)	Surface temp. °C (°F)
Io	1821 (1132)	421,600 (261,982)	1.8	3.53	8.9 (8.8)	−150 to +300 (−116 to +572)
Europa	1565 (972)	670,900 (416,897)	3.6	2.97	4.8 4.7	−220 to −130 (−186 to −96)

It is therefore probable, in such an energetic environment, that chemical reactions between simple molecules occur, leading to the formation of higher organic compounds. Unfortunately, rapid rotation of the planet and high radiation levels combined with vigorous movement of the atmosphere conspire to render the Jovian atmosphere an extremely unlikely environment in which life might take root. Even so, the chemical reactions presumably taking place within the atmosphere are of interest when considering parallels with reactions that might have occurred on the primordial Earth.

Jupiter has 16 satellites. The four largest and brightest (Io, Europa, Ganymede and Callisto), all similar in size to the Moon, were discovered in 1610 by Galileo Galilei (and are known as the Galilean satellites). Through observations over the past 360 years, astronomers have compiled some basic

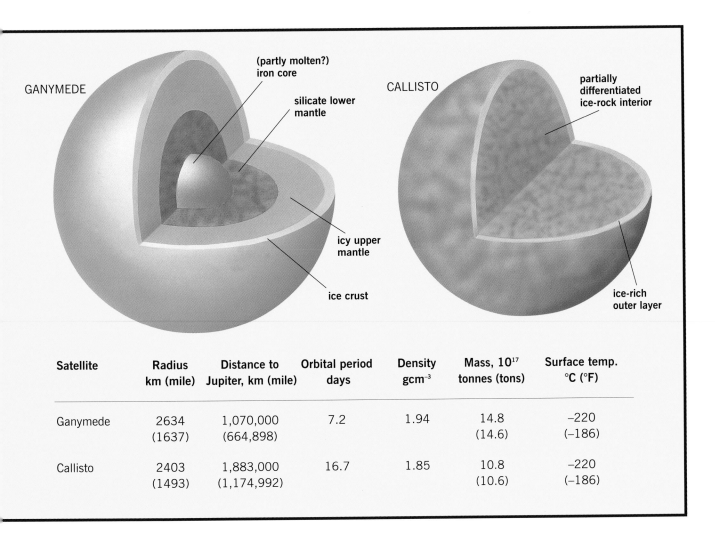

Satellite	Radius km (mile)	Distance to Jupiter, km (mile)	Orbital period days	Density gcm^{-3}	Mass, 10^{17} tonnes (tons)	Surface temp. °C (°F)
Ganymede	2634 (1637)	1,070,000 (664,898)	7.2	1.94	14.8 (14.6)	−220 (−186)
Callisto	2403 (1493)	1,883,000 (1,174,992)	16.7	1.85	10.8 (10.6)	−220 (−186)

information about the satellites. From these parameters we can deduce the rough composition of each satellite and it seems that they are very different from each other. Both the Voyager 1 and 2 missions of 1979, and the Galileo mission that started in 1995, provided spectacular images and a wealth of data for the Galilean satellites. It is clear that the four bodies exhibit a range of properties and environments that maintain the potential to harbour life.

Io

Io is the innermost of the Galilean satellites (but not the innermost satellite of all since there are four minor satellites orbiting closer to the planet). Its bulk density (see above) indicates that it is composed of metal and silicates. Recent gravitational and magnetic data from the Galileo mission suggest that Io has a dense core of iron and iron sulphide surrounded by a molten silicate interior overlain by a thin silicate crust. The diameter

ABOVE **Global view of Io, the most volcanic body in the Solar System. No impact sites can be seen, only volcanic craters and lava flows.**

Even though there is no indication that water is present – either at its surface as ice, or as water vapour, hydrogen or oxygen in its atmosphere – Io is included in the list of potential habitable bodies on the basis of its thermal properties. There are several sulphur-loving species on Earth that survive more than adequately in hot springs, around geysers and in volcanic craters, as we have seen. Perhaps similar organisms have carved out a niche on Io. Even so, any such organisms would have to be sufficiently hardy to survive the intense radiation from the clouds of energetic particles that flow along the magnetic field lines between Io and Jupiter.

Europa

Europa orbits Jupiter at a distance sufficiently close for the satellite to be heated by the tidal forces of Jupiter's gravitational attraction. Europa is potentially one of the most exciting Solar System bodies on which to search for traces of life. The reason for the interest in the satellite came from information and images acquired by NASA's Galileo mission of 1995 onwards. Europa's surface is one of the brightest in the Solar System; its density (see p. 62) implies that it is made of metal and silicates, but with a significant water-ice content. Gravity and magnetic data suggest that Europa has a metallic core, overlain by a silicate mantle, over which resides a crust of water-ice – it is sunlight reflecting from the ice that makes Europa so bright. Europa also appears to have a magnetic field.

The most recent models for Europa's 150 km (93 miles) thick icy crust depict a

of the core is about half that of the satellite. Io is heated by the tidal forces of Jupiter's gravitational attraction, a frictional force so great that Io exhibits substantial thermal activity. In fact, Io has one of the most energetic surfaces in the Solar System, with silicate lava flows from active volcanoes covered by sulphur deposits emanating from geysers and vents. No impact craters mark Io's face: the satellite is continually resurfaced by fresh lava flows. The surface temperature range is between –150°C and +300°C (between –116°F and +572°F), depending on locality and proximity to thermal vents.

layered structure, with a substantial sub-surface salt-rich ocean beneath a thinner shell of water-ice, with the salty ocean responsible for the magnetic field. If there is indeed a sub-surface liquid ocean on Europa, then the heat source that keeps it liquid is likely to come from the gravitational interactions between the orbiting Galilean satellites and Jupiter: both Io and Ganymede exert a minor gravitational pull on Europa, in addition to the much larger effect of the parent planet itself.

Europa's surface records a history of activity and change: as well as being marked by craters and associated impact ejecta, Europa is scarred and crossed with a complex sequence of ridges and grooves. The surface features are traced out by variations in brightness, and have been interpreted as variations in the composition of the material that forms them. Darker regions are probably richer in dust than the brighter, ice-rich areas.

Impacts excavate dust- and salt-rich material from below the icy crust, and distribute it across the surface. There may also be tectonic activity: not driven by molten silicate, as on Earth and Io, but driven by upwelling plumes of slushy water-ice, in a process known as 'cryovolcanism'. There may also be explosive activity, similar to geysers on Earth, erupting through cracks in the ice, again depositing dust across the surface.

Life on Europa?

So what is the probability that life has arisen on Europa? Are all the basic ingredients present? The structure of Europa's surface seems best explained by the presence of a liquid or semi-liquid water layer beneath a crust of ice. Europa has been cratered, and, like Earth, presumably received organic compounds from the comets and asteroids that bombarded it. The final requirement for life, that of an energy source, is also fulfilled on Europa.

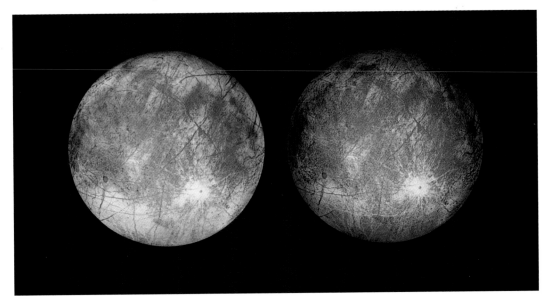

LEFT **Two global views of Europa, on the left in natural colour, on the right colour-enhanced. Brown is dust, ice is blue. The lines are fractures, up to 3000 km (1850 miles) in length. The impact crater Pwyll to the lower right of centre is 50 km (30 miles) across.**

RIGHT **Domes, ridges and blocks of ice in disturbed terrain on Europa. Area shown is 250 x 200 km (160 x 125 miles).**

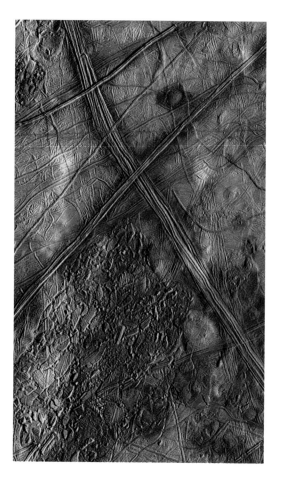

Sub-surface liquid water is maintained by heat generated from the gravitational tug between Jupiter (and Io) and Europa. There is also a theory that sufficient chemical energy could be produced from radiation to maintain microbial life at Europa's surface.

There has been much speculation that Europa's ocean might be heated from the bottom upwards, by hydrothermal vents similar to those found on the deep ocean floor of the Earth. If that is the case, then, also like the Earth, hydrothermal vents on Europa might host a rich variety of organisms.

Future exploration

Further exploration of Europa is well into the planning stage. Scientists hope to be able to verify the presence of a sub-surface water ocean using radar imaging. If an ocean is present, then the depth of the ocean will be determined, and any vertical structure (liquid and slush layers, for instance). The next step would then be to penetrate the ice and investigate the ocean and its floor.

RIGHT **Close-up of the Conamara region of northern Europa, showing disrupted rafts of thin ice. Area shown is 70 x 30 km (45 x 20 miles).**

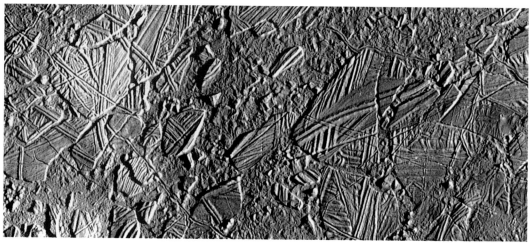

In the meantime, the Earth has an almost ideal test site for future exploration of Europa: Lake Vostok, a freshwater lake, located some 4 km (2.5 miles) below the ice of the Antarctic plateau. It was first detected by radar in 1974, and appears to have been isolated from the surface for possibly millions of years. Viable micro-organisms have been found in cores drilled through the ice down to a depth of around 2.4 km (1.5 miles). Lake Vostok provides a valuable opportunity for the potential investigation of primitive terrestrial micro-organisms that might have developed and evolved along a different path from their surface-located relatives.

Ganymede and Callisto

Ganymede and Callisto are the two largest of Jupiter's satellites and have similar densities (see p. 63). They are, however, different in appearance, due to their differing evolutionary histories. Even so, both satellites have properties that imply they might support life. Magnetic and gravitational data from the Galileo mission have allowed scientists to propose models for the internal structures of Ganymede and Callisto.

Ganymede appears to have a small iron core overlain first by a silicate and then by an icy mantle, encased in a crust of ice. Its surface is characterised by distinctive regions of bright and dark terrain. The latter is marked by numerous craters, indicating a relatively ancient age, and also by long sinuous furrows. In contrast, the bright terrain is much smoother, with fewer craters. Instead of furrows, sets of parallel ridges and grooves cut across the surface. Ganymede's

furrows and grooves have been generated by tectonic activity, but of a different style from that which produced the ridges and grooves on Europa. The features on Ganymede are thought to be formed by stretching of the crust, resulting in parallel faults, similar to rift valleys on Earth; cryovolcanism then flooded the valleys with ice-rich material. Like Europa and Io, Ganymede is influenced by Jupiter's gravitational heating. Ganymede also appears to have its own magnetic field, implying that its core is still partially molten. By analogy

LEFT **Close-up view of the surface of Ganymede, showing a narrow band of bright terrain about 15 km (10 miles) wide running through older, dark terrain.**

BELOW **Global view of Callisto, natural colour on the left, enhanced colour on the right. The bright structure at the lower right is the Valhalla impact area, about 600 km (360 miles) in diameter.**

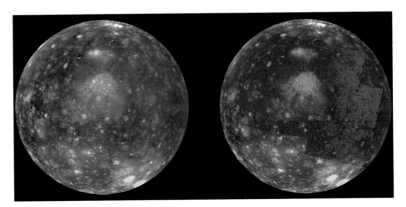

RIGHT **Global view of Ganymede, the largest satellite in the Solar System. The dark regions are older, heavily cratered terrain.**

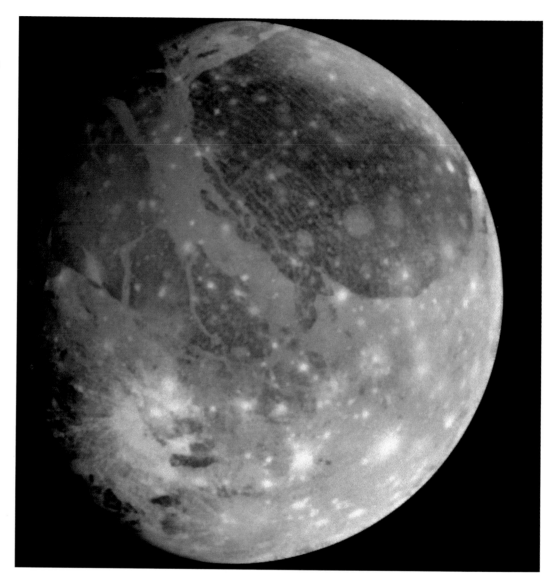

RIGHT **Global view of Ganymede, the largest satellite in the Solar System. The dark regions are older, heavily cratered terrain.**

with Europa, it is possible that at least part of Ganymede's icy mantle might yet be liquid, in which case, once again it might be possible that some type of life resides there.

Callisto presents a completely different appearance from Ganymede: it is a mixture of rock and ice, topped by an ice-rich crust. Its entire surface is pock-marked by craters and multi-ringed features, the largest of which is approximately 600 km (373 miles) in diameter and named Valhalla. Callisto does not have its own magnetic field, and thus no molten core – the satellite is sufficiently far from Jupiter that tidal heating has played no part in its evolution. However, the Galileo mission did record magnetic anomalies when it flew by Callisto, suggesting some kind of interaction with Jupiter's magnetic field. One interpretation of the measurements is that the magnetic field was induced by a salt-rich ocean; models suggest that such a global ocean on Callisto would be between 10 and 100 km (6 and 60 miles) deep. The heat source that has kept such a layer liquid is, as yet, unknown. Although it perhaps seems unlikely that life might have arisen on Callisto, the possibility cannot be ruled out entirely.

Jupiter satellites summary

To summarise the observations for the Galilean satellites of Jupiter: Io, Europa and Ganymede are all subject to the influence of Jupiter's gravitational pull. As a result of this, they are all heated and are all molten internally to varying degrees. Given, then, that these satellites are rocky and presumably have their primordial carbon inventories augmented by impacting comets and asteroids, and have a ready energy source, they each, again to varying degrees, show the potential to harbour primitive organisms. Europa is the one identified to have the best possibilities for life, mainly because it seems to possess a fairly deep ocean of liquid water. Future space missions will, no doubt, focus on Europa in any search for life.

Titan

The gas giant Saturn is easily recognised by its spectacular ring system composed of a myriad blocks of ice and rock. The complexity of the ring system, and the variety of properties shown by the planet's numerous satellites were recorded by the Voyager fly-by missions of 1980 and 1981. Titan, a mix of rock and ice, is Saturn's largest satellite and the second largest satellite in the Solar System (after Jupiter's Ganymede). It was a prime target for Voyager 1, since data from ground-based observations and the 1979 Pioneer 11 encounter, showed that Titan had an atmosphere of nitrogen, methane and ammonia. There were great hopes that Voyager 1's encounter with Titan would provide images of the surface of this enigmatic body. Unfortunately, Titan was covered with thick clouds which made it impossible to take any pictures of the surface. Even so, the instruments on board the

BELOW **A heavily cratered region near Callisto's equator. The prominent crater just left of centre is an impact crater about 20 km (12.5 miles) across.**

Voyager craft were able to gain data to build up a much better picture of the satellite.

Titan is enveloped by a thick atmosphere, the main component of which, like the Earth, is nitrogen. However, unlike the Earth, the remaining constituents are methane and argon, with trace amounts of higher hydrocarbons (such as ethane and propane), carbon monoxide, hydrogen cyanide and cyanogen.

Radar measurements from Earth, coupled with observations with the Hubble Space Telescope, show Titan as having a rough rocky and icy surface with a temperature of about −200°C (−166°F). Condensation of

RIGHT **Global view of Titan, Saturn's largest satellite, covered in clouds of hydrocarbon haze.**

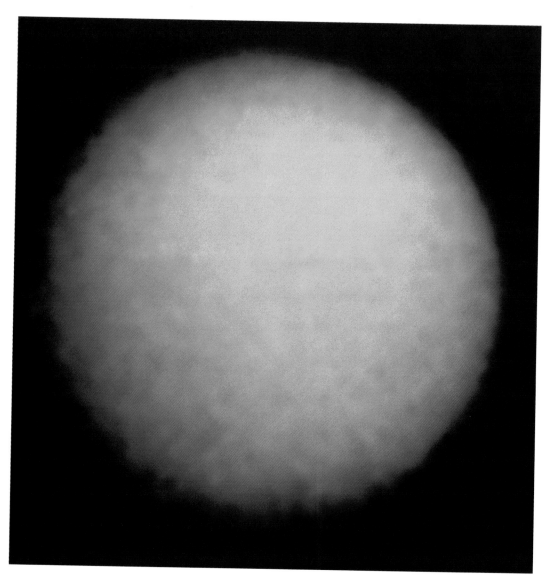

LEFT **Artist's impression of Titan's surface, showing the Huygens probe entering the atmosphere.**

ethane from the atmosphere has resulted in the build up of lakes and ponds of ethane in which ammonia and methane are dissolved. Although there is no suggestion that the environment of Titan is in any way suitable to harbour life, the atmospheric composition and surface conditions are sufficiently primitive that they may be similar to conditions found on the early Earth. Understanding the atmospheric and surface processes on Titan may help us to understand the processes that led to the evolution of life on Earth. For this reason, Titan is the target of Huygens, an ESA-led probe (part of the joint NASA-ESA Cassini-Huygens mission to the Saturnian system, scheduled to arrive in December 2004), which will descend through Titan's atmosphere. The probe will acquire images, measure the elemental and isotopic composition of the atmosphere and record its environmental conditions right down to its crash (or splash) landing on the surface.

OPPOSITE **A typical spiral galaxy, rich in dust as well as gas. Stars are clustered more densely at the centre of the galaxy than in the arms.**

RIGHT **Part of Hubble Deep Field showing a rich variety of galaxies, any one of which might contain thousands of stars capable of supporting a planetary system on which life might have developed.**

Our Solar System and beyond

A voyage through the Solar System, in search of water and energy at levels appropriate for life to originate, and possibly even survive, reveals a surprising diversity of habitats – from the dry soils of Mars to the sulphurous geysers of Io, the ocean of Europa and the ethane and ammonia lakes of Titan. Several space missions are planned for the coming decade, which may answer some of the issues raised here. But until then we can only speculate on the presence, or otherwise, of micro-organisms on bodies other than the Earth within our close neighbourhood. If we are going to speculate, then we need not confine ourselves to the Solar System. Ours is but one star, in a galaxy of a thousand million stars. What would a search for life beyond our Solar System reveal?

Life beyond the Solar System

The study of astrobiology is not confined to the Solar System, but extends to the search for planets orbiting other stars and the potential for life on those planets. The concept of a Habitable Zone (HZ), as discussed earlier, can be extended beyond our Solar System. The Sun is in the HZ of the Milky Way – perhaps the Milky Way is in the HZ of the Universe.

If we allow the search for life to be governed by conditions found on the Earth, then we must look also at what sets our star apart from other stars. Our Sun is a hydrogen-burning main sequence star, around 4.56 billion years old. But what other types of star are there, and do they have the potential to host a planetary system amongst which is a planet on which liquid water is stable?

Stars

Stars are classified according to their temperature (which governs how brightly they shine) and size. Only a few types are likely to support planetary systems. For a start, possession of a planetary system implies the presence of dust – or at least some type of solid material. As we have discussed, solid material comes from nucleosynthesis within stars – the burning of hydrogen to helium and eventually to carbon, oxygen, silicon, magnesium and iron. Without several generations of stellar cycling to produce these elements, no planets could form. Star systems made up of the very oldest stars, such as

globular clusters, have very low amounts of elements other than hydrogen and helium, and so are unlikely places to look for planets.

Another very important facet of our star's suitability as a parent is its lifetime, and its stability over an extended period, allowing time for life not just to arise (which appears to have happened remarkably rapidly on the Earth), but to evolve into the vast biodiversity that is seen today. Highly energetic stars that

OPPOSITE **The Sun, as captured by the SOHO spacecraft.**

BELOW **The bright ball in the centre of the picture is a globular cluster of around 300,000 stars. It is located within the Andromeda galaxy.**

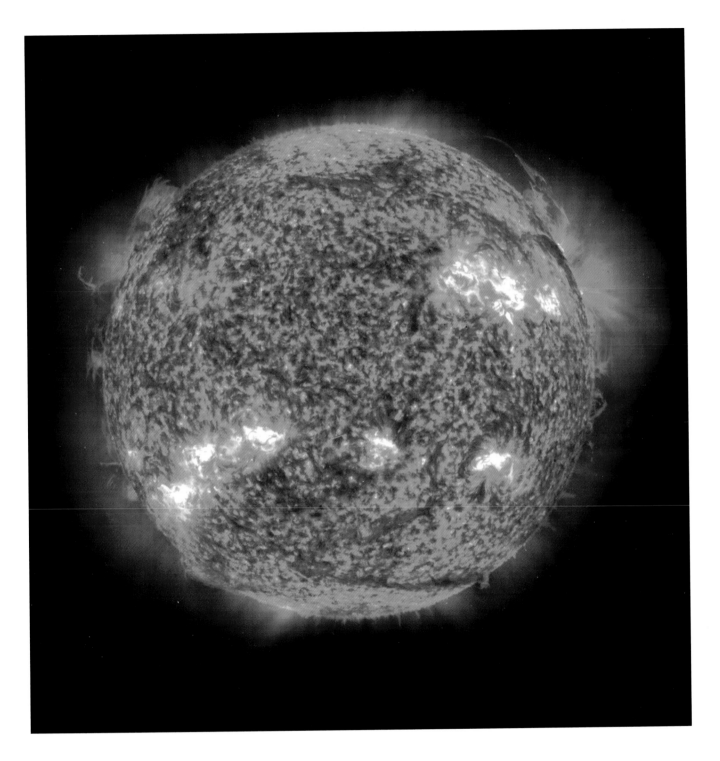

are susceptible to sudden outbursts of radiation, or explosion, or extremely rapid spin are much less likely to provide extended periods of stable conditions for the uninterrupted development of life. Into this category fall giant stars that are liable to become supernovae or novae as they reach the end of their lifetimes, and rapidly rotating stars called pulsars.

The lifetime of a star is related to its mass: those that are more massive than the Sun have a shorter lifetime, say one to two billion years rather than the 10 billion years of our Sun. So more massive stars will probably not be stable for long enough for higher types of life to evolve. Additionally, massive stars emit much more ultraviolet radiation than stars like our Sun: higher amounts of uv radiation can lead to stripping of the atmosphere followed by destruction of organic molecules on the planet's surface. Massive stars are therefore unlikely to be hosts for life-bearing planetary systems.

Stars less massive than the Sun also have their share of problems. They are cooler, emitting less heat. So for a planet to be within the Habitable Zone of such a star, it would have to orbit close to the star, resulting in problems from higher gravitational attraction. So, although there are billions of stars in our Galaxy, many of them are types (too old, too big, too small, too energetic) that are unlikely to support planetary systems for sufficient lengths of time for lifeforms to arise or evolve.

Galaxies

The same arguments apply when we look at galaxies, which are vast accumulations of stars. Galaxies also evolve and change with time: they collide and merge with each other, cannibalising dust and gas from their companions in the process. The oldest galaxies that we can observe have been imaged by the Hubble Space Telescope. The distribution of galaxies within the Universe has been determined by astronomical observation programmes, such as that undertaken at the Anglo-Australian Observatory (the 2dF survey), which aims to map the distribution of some 250,000 galaxies within a particualr region of the sky. So far the survey has located about 100,000 galaxies. The maps show that there is an overall pattern to the distribution of galaxies within the Universe: they are concentrated in sheets and ribbons, with voids between them. Theoretical modelling of galactic distribution gives a similar result. This structure relates directly to the processes that occurred during the period of rapid expansion just after the Big Bang. Are there more favourable regions

BELOW **Map of a 'slice' of the Universe showing the distribution in space of over 100,000 galaxies, measured by the Anglo-Australian Telescope's 2dF instrument. It shows that galaxies are clustered into clumps, with voids in between. This structure comes from the way in which matter and energy were spread out after the Big Bang.**

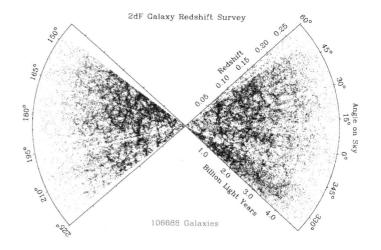

RIGHT **Two galaxies colliding: NGC 2207 on the left, IC 2163 on the right.**

RIGHT **Two galaxies colliding: NGC 2207 on the left, IC 2163 on the right.**

within the Universe where life might be more likely to arise?

If the type of star is important for life, then so too is the location of the star within a galaxy, and also the type of galaxy. Dust-poor elliptical galaxies are likely to be depleted in the elements necessary for planet building.

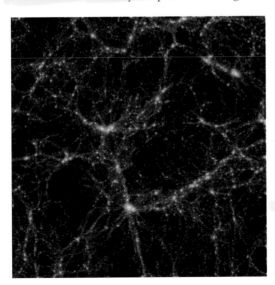

RIGHT **An experimental simulation of how particles clump together to form galaxies. By varying the starting conditions of the particles, such as their density and energy, cosmologists can try and simulate patterns of voids and clumps that match the observed structure of the Universe, and so learn more about how the Universe formed.**

Spiral galaxies (like the Milky Way) have plenty of dust mixed in with the gas, and so have the potential for planet formation. Even so, within galaxies there are regions where conditions for life are not favourable. Towards the central hub of a spiral galaxy, the number of stars increases, leading to decreased distances between stars, and an increasing number of neighbours per star. When stars are close together, their mutual attractions will be increased, resulting in higher gravitational instabilities for the interacting stars.

Greater densities of stars means that there will be higher radiation pressures and stellar winds within the central region of a galaxy than the outer fringes, plus a higher rate of stars becoming novae and supernovae, with accompanying explosions of energy. When we look at this picture, we see that the Sun is in the HZ of the Milky Way: it is far out enough to be relatively isolated from its nearest

Jazz

Searching for Earth-like planets

Both ESA and NASA are planning missions to search for Earth-like planets orbiting nearby stars, to take spectral images of the stars and decide on the potential for life on any of the planets detected. The Darwin project has been selected by ESA as one of its forthcoming Horizon 2000 space missions, for launch in 2012, whilst NASA's Terrestrial Planet Finder (TPF) is still under consideration for launch in the next decade. It is possible, given that the goals and timescales of the missions are so similar, that a joint mission will be planned.

Darwin is a suite of five or six telescopes, each 1 m (3 ft) in diameter, clustered around a central hub, whilst the TPF is currently proposed as four telescopes each 3.5 m (10.5 ft) in diameter, also around a central hub. Each set of telescopes form a network of receivers which simultaneously detect light from a star. Combining the signal for each receiver allows detection of the faint light reflected from any objects close to the star (such as planets).

The telescopes operate in the near infra-red end of the spectrum, a region that is more appropriate for comparison with the Earth and rocky planets, as signals from dust can be detected. The telescopes can also perform high resolution spectral imaging of the star.

ABOVE **Artist's impression of what the suite of telescopes might look like in the search for an Earth-like planet.**

neighbours and in less danger from the effects of potential supernovae, but it is not on the fringes of the galaxy, where the lower rate of star formation makes it unlikely that sufficient amounts of heavier elements (carbon, oxygen, silicon, etc.) have been produced to allow planet-building to take place.

Clusters of galaxies will also be regions of high activity – interaction between stars, turbulence, high levels of uv radiation, frequent supernovae and so on. Galaxies were clustered more closely together at an earlier time within the history of the Universe, and consisted of less-evolved stars. The evolution of galaxies, as well as the evolution of stars, must have played an important role leading up to the formation of life.

Searching for planets around stars

The search for planetary systems around stars, then, should be directed towards looking for Earth-like planets (ELP) orbiting Sun-type stars at distances considered to be within the HZ of the observed star (which is located within the HZ of its galaxy). The techniques and instrumentation needed to observe planets outside our Solar System have been developed relatively recently. Early success came in 1992, when three approximately Earth-sized planets were detected orbiting pulsar PSR 1257+12. Pulsars, or neutron stars, are the rapidly spinning remnants from supernovae and unfortunately, they are rotating far too fast to allow the possibility of life on any orbiting planets.

The first observation of a planet orbiting a Sun-type star was made in 1995, when a Jupiter-sized object was detected orbiting the star 51 Pegasi. Since then, more than 50 giant planets have been detected around nearby Sun-type stars, and more are detected each month. The main technique employed to make these observations is to look for

LEFT **An artist's impression of how an Earth-like planet orbiting a Sun-type star in another stellar system might appear.**

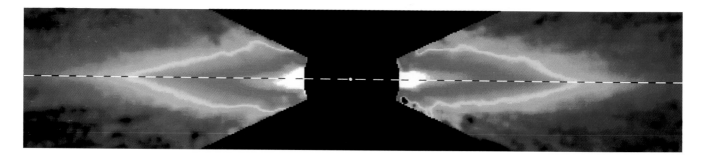

variations in the star's motion as an indirect indicator of a planet's presence; one planet has also been detected directly, from measurements of reflected starlight. So far, none of the planets discovered has had the characteristics that would designate it as a likely candidate for further study as the host for life: the planets have all been Jupiter-like gas giants orbiting close to their star, much closer than Mercury is to the Sun in our own planetary system.

The next stage in searching for extra-solar planets can only come with improvements in techniques that will allow the detection of Earth-like planets orbiting Sun-type stars. Both the European Space Agency (ESA) and NASA have missions in the planning stage (Darwin and Terrestrial Planet Finder, respectively), which will have the capability both to detect ELPs, and to determine their characteristics (see p. 78).

Dusty disks around stars

Another way to search for planets around stars is to look for the dusty disks that exist prior to the formation of individual planets. The advantage of this technique over planet observation is that disks can readily be found by looking out for the emission of excess infra-red radiation from a star, which results from the presence of dust grains. The Infra-Red Astronomical Satellite (IRAS) made the first observation of an infra-red excess around the star Beta Pictoris (β Pic).

This star is approximately 60 light years (around 600 million million km or 372 million million miles) away from the Sun. Subsequent optical imaging has produced a picture of a flat disk of dust extending away from the central star. The disk is about 900 AU across (where 1 AU is the distance from the Earth to the Sun), but the dust does not extend all the way to the centre. Images obtained by the Hubble Space Telescope show that the central inner zone (a region about 50 AU across, wider than Pluto's orbit, which stretches to 30 AU) is relatively free from dust, implying that the dust has perhaps been swept up into a planetary system. Although interpretation of the dusty disk around β Pic is not yet fully complete, to observe processes taking place in a protoplanetary disk around a star is an important step in constructing a history of our own star and planetary system.

Observation of infra-red excesses from dusty disks around stars is also a useful measure of how often potential life-supporting planetary systems form.

ABOVE **False-colour image of a dusty disk around the star β Pictoris. The disk is the type of region in which planets might eventually form. The disk is 320 billion km (200 billion miles) across, and the star is 50 light years away.**

How should we search for life?

The signature of life on Earth is indisputable; any spacecraft passing by our planet would detect evidence for life (abundant water and organic material), even evidence of intelligence (regular thermal and radio emissions). Unfortunately, such clear indications of life on other Solar System bodies is very unlikely. What should any mission to detect extraterrestrial life be looking for? It is important to define what we are searching for before we spend too much time looking in the wrong place for the wrong signals. As Carl Sagan, a pioneering astrobiologist, remarked in 1994:

'The search for extraterrestrial life must begin with the question of what we mean by life. "I'll know it when I see it" is an insufficient answer'.

Signatures of life might be trace gases in an atmosphere, or components in surface layers or in sub-surface regions; all possible niches must be examined. For life on Earth to

LEFT **Radar image of the Great Wall of China, 700 km (430 miles) west of Beijing, taken from the space shuttle Endeavour in 1994.**

BELOW LEFT **Satellite image of the Earth's surface, showing lights from centres of population, and rural fires. Images like this send a clear signal that the Earth is inhabited by a technological civilsation.**

succeed, the importance of water and organic compounds has been stressed; any search for extraterrestrial life is therefore likely to include, at the very least, the search for water and carbon. The relevant signatures fall into two groups: those that can be measured directly, and those that are observed remotely.

Direct signatures of extraterrestrial life

Measurements of organic compounds are likely to be of great significance when searching for extraterrestrial life. The simple presence of organic molecules is, of course, insufficient evidence to infer a lifeform, since organic compounds can have a purely non-biological origin (they are found in comets, for example). There are, however, several analytical techniques that can be used to determine whether organic molecules are by-products of biological processes.

The first method measures the 'optical chirality' of organic molecules (see opposite page). Some molecules exist in two versions (one a mirror image of the other). If there is more of one form of the molecule relative to the other form, then it is possible that the molecule was produced biologically, since nonbiological processes tend to result in equal mixtures of the two forms.

The second method for determining the biological, or otherwise, nature of extraterrestrial organic compounds uses the isotopic composition of the material as a marker. As mentioned before, atoms of the same element with slightly different masses are termed isotopes. Nonbiological systems tend towards a balance between the isotopes present in different compounds, but biology introduces disequilibrium. So when searching for life, say on Mars, we would look out for organic molecules (which are potentially biological in origin) that have different signatures from the inorganic carbon present in the soil, and from the carbon dioxide in Mars's atmosphere.

One of the key aims of the Beagle 2 lander, due for launch in 2003 on board ESA's Mars Express, is to search for chemical traces of life on Mars, by looking for an unbalanced isotopic signature between carbon in different samples. One of the instruments on board Beagle 2 will be a 'mole', a sampling device that glides across the surface of the planet and can burrow into soil or gravel, and collect material for analysis by the gas analysis package (GAP) on the lander platform. GAP

BELOW **The Beagle 2 lander, in its operational configuration, with solar panels opened out to show the base station where instruments and cameras are located.**

Optical chirality

Molecules that exist in two versions, one the mirror image of the other, like the two hands of a pair of gloves, are known as stereoisomers. They have identical physical properties, apart from the way in which they interact with a beam of polarised light: if the light is rotated clockwise, the molecule is termed dextrorotatory (D); if it is rotated anti-clockwise, it is termed laevorotatory (L). Nonbiological organic compounds created by chemical reactions are almost always produced as an equal mix of D and L forms. But when the same compound is produced biologically on the Earth it is present as only the L form. It is possible that a similar effect might occur with extraterrestrial biology.

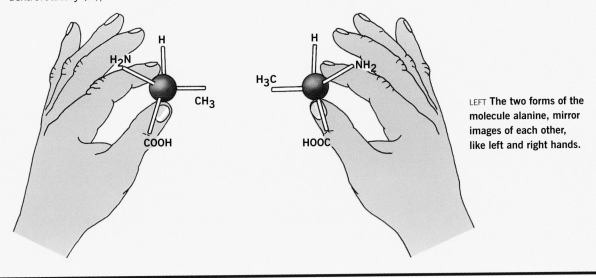

LEFT **The two forms of the molecule alanine, mirror images of each other, like left and right hands.**

will determine the amount, type and isotopic fingerprint of organic and inorganic carbon (such as carbonate minerals) in the soils, rocks and atmosphere, to build up a picture of a carbon cycle on Mars.

Another method for detecting biologically significant molecules in situ is to use the technique of Raman spectroscopy. Each molecule or class of compounds has a specific fingerprint, e.g. compounds associated with chlorophyll from photosynthetic organisms. If we detected such fingerprints in an extraterrestrial location, such as the surface of Mars, it would be very strong evidence for some type of biological activity.

Remote signatures of extraterrestrial life

Fortunately, since it is not always possible to analyse planetary materials directly, there are ways of searching for likely habitats for life from afar. The most common technique is to look at the heat and light reflected from a planetary body (its spectral signature), most usually of its atmosphere (see p. 85), which indicates which molecules are present.

As already mentioned, nonbiological systems tend towards equilibrium, and disequilibrium products, if present, can be taken to infer the presence of life. The presence of ozone (O_3), which implies the presence of oxygen, would indicate life.

In a lifeless system, oxygen and ozone are both removed very rapidly from the atmosphere by breakdown of the molecules by uv radiation, by combination with sediments, and by oxidation of volcanic gases. The dominant oxygen-producing mechanism that we know of is photosynthesis (nonbiological processes such as the breakdown of water in the upper atmosphere produce only minor amounts of oxygen). And so detection of ozone in a planetary atmosphere can probably be taken to imply the presence of life, although the converse is not true. Absence of oxygen or ozone does not necessarily imply the absence of life: there was no oxygen in the Earth's atmosphere for the first two billion years of its history, when there were plenty of micro-organisms flourishing in the waters and sediments, and the atmosphere was a reducing one.

Fermi's Paradox: Where are they?

In 1950, Enrico Fermi, one of the foremost physicists of the 20th century, when discussing the supposed existence of extraterrestrial intelligences (ETI), casually posed the question, "If there are extraterrestrials, where are they?" He rationalised that ETI could not exist, on the basis that they had not visited the Earth. His argument, which has since become known as Fermi's Paradox, is based on the following logic, regarding assumed lifetimes for technological civilisations to evolve.

Assume that a technological civilisation has evolved on a planet around a star, and assume that it takes approximately 1000 years (30 or so generations) for the civilisation to develop interstellar travel at around one tenth of the speed of light. Then spacecraft could leave the planet and travel to neighbouring stars. Even if it took 1000 years to reach stars with suitable planetary hosts, it would still only be a matter of a few generations before the planet was colonised, and additional spacecraft again sent outwards to the stars. And so, the argument went, it would only take a matter of a few tens of millions of years before the entire Galaxy was explored. Given the Galaxy is billions of years old and it is perhaps unlikely that the Earth is the first planet on which a technological civilisation should arise, then the lack of ETI visiting the Earth implies that there are none.

Arguments against this paradox can lead quickly into debates about the reality or otherwise of UFOs. Reasons for continuing the SETI programme include arguments that ETI might already have visited the Earth and left without leaving any signs of their presence, or might be monitoring humanity's progress towards interstellar travel. While such speculations provide a rich seam of material for science fiction writers, we must always bear in mind that one day, possibly even in the not too distant future, we might have to face the question of ETI more directly.

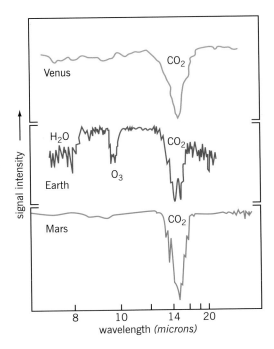

RIGHT **Comparison between the spectral signatures of Venus, Earth and Mars. Earth is the only one to show the signal for oxygen (as ozone).**

Detection of methane in an atmosphere is another clue that biological processes might be occuring. Methane is produced on Earth in different biological environments, e.g. by termites, and by rice in paddy fields. It is rapidly oxidised and thus removed from the atmosphere, so its presence is indicative of continual renewal from a biological source. If ozone and methane are present together it is almost certain that they have been produced biologically. Spectral techniques were tested when the Galileo spacecraft flew by the Earth and found evidence for life.

Future searches for Earth-like planets (ELPs) will concentrate on the detection of conditions that would indicate environments conducive to the survival of life. Infra-red spectrometers will be employed to measure

RIGHT **Paddy field landscape, a major source of methane to the Earth's atmosphere.**

the thermal emission characteristics of ELPs, to determine surface temperatures, and the presence or otherwise of water.

SETI – The search for extraterrestrial intelligence

So far this book has considered the search for life, and mostly fairly primitive life at that, rather than the search for intelligent life. However, the last 20 years have been marked by programmes to search for intelligent life (the search for extraterrestrial intelligence, or SETI). The main thrust of SETI has been to monitor incoming radiowaves from stars, to look for patterns in the frequencies that might indicate an artificial signal – one sent deliberately by an extraterrestrial intelligence

– as opposed to the regular, but natural, signals that emanate from pulsars. The main telescope used by SETI is the Arecibo telescope in Puerto Rico, but the UK's Lovell telescope at Jodrell Bank in Cheshire is also an important contributor.

One of the key instigators of SETI has been the US scientist Frank Drake. He formulated an expression (the Drake equation, see below) to estimate the number of civilisations that might be present in the Milky Way Galaxy.

The quantity of data acquired from continually scanning the sky across a range of wavelengths is enormous, and the SETI programme was in danger of being overwhelmed, until the novel proposal of

The Drake Equation

The Drake equation was formulated by Frank Drake in 1961, to estimate the number of communicating civilisations N that might be present in our Galaxy, the Milky Way:

$$N = N^* \times f_s \times f_p \times n_e \times f_l \times f_i \times f_c \times f_L$$

Where:

N^*	=	Number of stars in the Milky Way
f_s	=	Fraction of Sun-type stars
f_p	=	Fraction of Sun-type stars with planets
n_e	=	Number of planets in the star's Habitable Zone
f_l	=	Fraction of habitable planets where life arises
f_i	=	Fraction inhabited by intelligent beings
f_c	=	Fraction inhabited by a civilisation that communicates
f_L	=	Fraction of the lifetime of a planet that it has an intelligent, communicating civilisation

Although there is no 'correct' answer to the equation, by assigning realistic limits for the various parameters, a range of probabilities can be calculated for different scenarios. As more information becomes available, and we reach a greater understanding of the constraints on life evolving, then the expression becomes more prescriptive. It is a truism that the value for N is at least 1, given that intelligent life has been detected on the Earth.

SETI@home was developed. The principle behind it is that the processing required to reduce the observed data is fairly straight-forward, and needed either one very powerful computer to process all the data, or many smaller computers (such as desktop PCs), each processing small packets of data. SETI@home allows anybody with access to the Internet to take part in SETI; the programme, launched in 1999 has proved

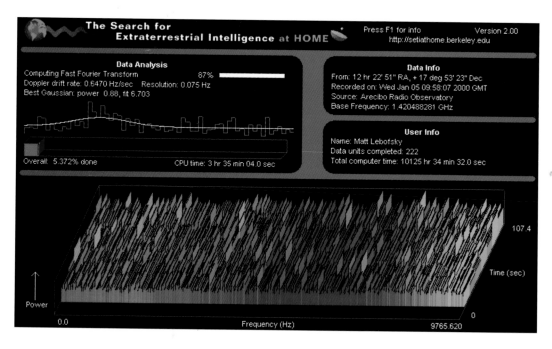

LEFT **The screensaver for the SETI@Home project, which uses the idle time from desk-top computers all over the world to process data received at the Arecibo telescope.**

BELOW **Illustration of the message sent out from the Arecibo telescope in 1974.**

even more successful than its organisers anticipated, and now over two million people worldwide take part.

Communication is not a one-way process, and much debate has centred around proposals to beam signals out from Earth to space, to make our 'voice' heard to any listening intelligences. The Earth emits signals every day, from TV and radio broadcasts, satellite communications and heat and light from population centres. However, all of these signals are weak and non-directional. Scientists have made several attempts to send specific messages out of the Solar System.

In 1974, the SETI team directed a powerful signal from the Arecibo telescope in Puerto Rico to the globular cluster M31, about 21,000 light years away. The message was a digital stream of data, which, when decoded, gives simple pictures of the Solar

System, DNA, a man and a radio telescope (see right). Advances in understanding have since shown that stars in globular clusters are unlikely to be hosts of extraterrestrial life. In 1999, a second signal with similar content to the first was sent from a radio telescope in the Ukraine, this time targeted towards a group of four stars approximately 60 light years away.

An even more speculative and problematic approach to interstellar communication was taken by scientists working on the Voyager programme. The two Voyager spacecraft were launched two weeks apart in late summer 1977, to explore the outer planets of the Solar System. Between 1979 and 1989 they returned spectacular images of Jupiter, Saturn, Uranus, Neptune and their satellites, before heading outwards beyond Pluto. On board each of the craft was a disk bearing a message

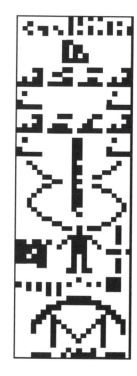

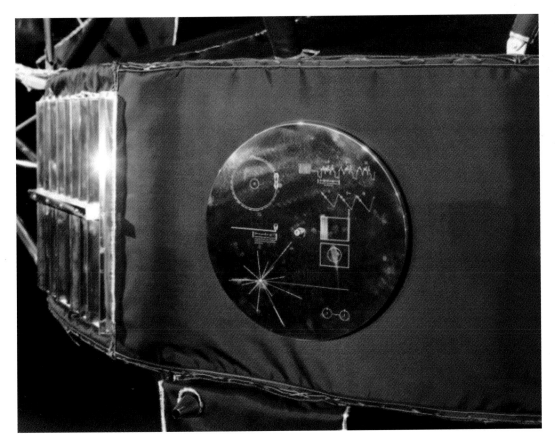

LEFT **The disk on board the Voyager spacecraft, bearing a series of images and sounds as a message from humanity.**

from humanity: a series of images and sounds recorded on a 12 inch gold-plated copper disk, similar to a long playing record. The images were selected to represent the diversity of life and culture on the Earth. Voyager 1 is still heading away from the Sun, but has not yet reached the outer limit of the Solar System. Voyager 2 is on a different trajectory, also heading away from the Sun. It will be decades before these spacecraft go beyond the Oort Cloud that marks the boundary of the Solar System and travel into interstellar space.

Epilogue

'Life' was earlier defined in terms of the ability of an entity to grow, to reproduce and to evolve. Although we have considered the conditions necessary for life to arise in an extraterrestrial habitat, the ability for life to evolve is perhaps much less easy to assess. Based on the range of conditions in which microbes exist and flourish on Earth, it seems fairly clear that there are many niches throughout the Solar System that could provide suitable habitats in which life might emerge. There is, however, a distinction that must be made between the emergence of life and its successful evolution. The fossil record indicates that life on Earth started up the evolutionary pathway only a matter of a few hundred million years after the formation of the planet. Nonetheless, it has been suggested that primitive life might have started more than once during this first epoch, only to be eliminated during periods of bombardment by asteroids and comets. It was not until conditions stabilised, at around four billion years ago, that the evolutionary process got underway.

Life on Earth has evolved through a series of stages, millennia of gradual, progressive evolution punctuated by sudden, catastrophic extinctions, after which new or different species diversified and became successful. The most obvious example of a change in evolutionary direction occurred at the end of the Cretaceous period, some 65 million years ago, when the dominant animal group, the dinosaurs, became extinct. The trigger that is generally believed to have caused this extinction was the impact of an enormous body, about 10 km (6.2 miles) in diameter, at Chicxulub, just off the Yucatan peninsula in the Gulf of Mexico. The environmental changes that followed from the impact included rapid swings in temperature, from very hot (as energy from the impact shock was transferred around the globe as radiant heat, setting fire to forests) to very cold (as the dust from the impact, coupled with smoke from the forest fires darkened the skies, cutting out light and heat from the Sun). Among the survivors of the disaster were a small group of fur-covered creatures, some of the earliest mammals, that eventually evolved into primates, and thus humans.

Although the impact at the end of the Cretaceous is the one about which most has been written, there have been several other global extinctions throughout the geological record. These events have not been coupled with asteroid impact, but with other mechanisms such as global changes in sea level or enhanced volcanic activity, and have also resulted in major changes in the developmental pathway of organisms. So both internal and external agents are responsible for evolutionary change on the Earth. Could similar events affect other bodies in the Solar System?

Looking back at the evolution of life on Earth from an astronomical and geological perspective, as has been discussed, we can see that many independent factors have conspired to produce the diversity of organisms present on the Earth today. The position of the Earth with respect to the Sun; the presence of the Moon and Jupiter; the action of plate tectonics; the evolution of the atmosphere; the carbon cycle – all are highly specific to the Earth, whereas bombardment by asteroids and comets affects all bodies within the Solar System, and presumably planets orbiting around other stars. The evolution of life on Earth records variations that result from environmental changes, or stresses, combined with chance events (such as catastrophic impacts). Even though there has been a Solar System-wide delivery mechanism (via comets and asteroids) for the building blocks of life, and there are many environments suitable on other planets and satellites for life to emerge, different stresses and chances acting on the same starting materials would presumably result in the evolution of different types of life from that on Earth. It is far from clear that

there has been any emergence or evolution of life on other bodies, which is why there is currently so much interest in astrobiology and the planning of appropriate space missions to investigate and explore other worlds.

So are we alone?

The discovery of micro-organisms on Earth that seem to be able to overcome the disadvantages of extreme heat, cold, pressure and radiation encourages the view that there may be life elsewhere. There is an abundance of organic materials and water within the Solar System, and the potential for their presence on planets around other stars. But even if microbial life is widespread within the Solar System or galaxy, there is no guarantee that it has managed to survive and evolve into sentient or intelligent beings. If results from missions such as ESA's Mars Express and NASA's Mars 2003 are as inconclusive as those of the Viking biology experiments, then perhaps, although we might not be alone, we do not yet have anyone with whom we can talk.

Glossary

asteroid – a small body (rocky, iron-rich or carbonaceous) orbiting the Sun in a near-circular orbit mostly between Mars and Jupiter

astrobiology – the study of the origin, evolution, adaptation and distribution of past and present life in the Universe

astronomical unit – the mean distance between the Earth and the Sun; approximately 150 million km (93 million miles)

Big Bang – a widely accepted model for how the universe formed approximately 15 billion years ago, from the explosion of a point of infinite density, pressure and temperature

biogenic – relating to living organisms (*c.f. organic*)

black hole – a region of space where matter is so dense that its gravitational pull prevents light escaping. Can be produced when massive stars collapse

CAI – calcium, aluminium-rich inclusions, found in chondrite meteorites, the oldest solids formed within the Solar System

carbon cycle – the circulation of carbon (as carbon dioxide) between living organisms and the Earth (atmosphere, water and rock)

chemosynthesis – the use, by simple organisms, of energy from chemical reactions to build up complex organic compounds (*c.f. photosynthesis*)

chirality – two versions of the same molecule, identical in all respects except that their structures are mirror images of each other, like left and right hands

chondrite – a stony meteorite that has not melted since its aggregation early in the history of the Solar System; most chondrites contain chondrules

chondrule – a small, millimetre-sized spherule of silicates rapidly cooled from a melt

comet – a primitive body of ice and dust that orbits the Sun in an elliptical orbit

core – the central part of a planet or satellite; may be molten or solid

cosmic dust – small particles < 1 mm in size produced from asteroids and comets; also known as *micrometeorites* or *interplanetary dust*

crust – the outermost solid layer of a planet or satellite

cryovolcanism – eruption of cold material (slush or ice) onto a planetary or satellite surface

DNA – deoxyribonucleic acid, the complex molecules present in all chromosomes that carry the code for hereditary characteristics

ecliptic plane – the approximate surface in which planets orbit the Sun

electron – an elementary particle, with a single negative electric charge, that orbits the nucleus of an atom

fusion – the reaction between two atomic nuclei, resulting in the formation of a single, heavier nucleus, with accompanying release of energy

galaxy – a large group of stars (around 100 billion) held together by mutual gravitational attraction

gravitational attraction – the force that acts between different bodies as a result of their mass

Habitable Zone – the region within a planetary or stellar system where conditions are sufficiently suitable and stable for life to arise

Herzsprung-Russell diagram – a plot of the luminosity of a star against its temperature showing the main evolutionary stages of different types of star

Hubble Space Telescope – a 2.4 m telescope, launched in April 1990 by ESA and NASA, orbiting the Earth at an altitude of approximately 600 km

hydrogen burning – the reaction that takes place in the cores of main sequence stars, where hydrogen nuclei are fused together to produce helium

hydrothermal vent – a break in the Earth's crust where hot, mineral-rich fluid escapes; also known as a *black smoker*

interstellar medium – diffuse material between the stars in a galaxy

isotopes – atoms of an element that have the same number of protons and electrons, but differing numbers of neutrons (and therefore different atomic masses)

Kuiper Belt – a belt of small rocky and icy bodies that orbits the Sun beyond Neptune

light year – the distance light travels through space in a year, approximately ten million million km at a speed of 300,000 km per second

luminosity – the brightness of a star

main sequence star – a star in which hydrogen-burning is occurring

mantle – the layer above the core, but below the crust of a planet

meteorite – a natural object that survives its fall to Earth from space

mineral – a natural substance with a definite chemical composition; rocks are mixtures of minerals

nebula – a cloud of gas and dust in which stars form

neutron – an elementary particle, with no electric charge, which is present in all atomic nuclei (except hydrogen)

neutron star – a star that has collapsed in on itself, and consists almost entirely of neutrons (*pulsar*)

nova – a star that suddenly increases in brightness, then gradually dims

nucleosynthesis – the formation of elements by reactions between atomic nuclei

nucleus (atomic) – the central region of an atom made up of neutrons and protons

nucleus (cell) – the part of an individual cell that contains the chromosomes that control hereditary characteristics

obliquity – the angle of tilt of a planet's axis relative to the ecliptic plane; the value is around 23° for the Earth

orbit – the path of one body round another as a result of their mutual gravitational attraction

organic – relating to molecules containing carbon and hydrogen (*c.f. biogenic*)

photosynthesis – the use, by plants, of energy from sunlight to build up complex organic compounds (*c.f. chemosynthesis*)

planetary nebula – an accelerating cloud of

material ejected from a star in its final stages of evolution

plate tectonics – the movement of rigid rafts of Earth's crust, away from spreading centres, where new crust is produced, towards subduction zones, where crust is recycled into the mantle, or to collision zones, where crust is uplifted into mountains

proton – an elementary particle, with a single positive charge, that is present in all atomic nuclei

protoplanetary disk – the cloud of gas and dust that surrounds a newly formed star, and that has the potential to evolve into a planetary system; sometimes abbreviated to 'proplyd'

pulsar – a rotating neutron star with a diameter of about 10 km, that emits radio waves with a regular frequency

radiation – emission of energy from a source

radioactive decay – the breakdown of the unstable isotope of an element into stable isotopes of different elements by emission of sub-atomic particles, with an associated release of energy

red giant – an evolved star that has burnt all the hydrogen in its core to helium, then expanded in size and cooled

satellite – a body which orbits another body of greater mass under the result of their mutual gravitational attraction, e.g. the Moon is the Earth's satellite

SETI – the Search for Extraterrestrial Intelligence, carried out by monitoring incoming signals from space using radio telescopes

solar wind – the stream of particles that flow outwards from the Sun

solfatara – a volcanic vent from which gases are emitted

sputtering – the process of alteration by bombardment by atoms or ions

spectrum – the pattern produced by separation of signals on the basis of wavelength, frequency, etc.

stone – a meteorite mainly composed of silicate minerals with additional metal and sulphides

supernova – a sudden, rapid explosion of a star, ejecting matter and energy into interstellar space

white dwarf – a star at an advanced stage of its evolution, formed by collapse after it has exhausted all its fuel for nuclear fusion

Index *Italics refer to captions or labels which are separate to main text references.*

Further information

Further reading

Before the Beginning: Our universe and others, Martin Rees. Touchstone, 1997.

Earth: Evolution of a Habitable World, Jonathon I. Lunine. Cambridge University Press, 1999.

The Molecular Origins of Life: Assembling Pieces of the Puzzle, André Brack (Editor). Cambridge University Press, 1998.

The New Solar System, J. Kelly Beatty, Carolyn Collins Peterson and Andrew Chaikin (Editors). Cambridge University Press, 1999.

The Origins of Life, John Maynard Smith and Eörs Szathmáry. Oxford University Press, 1999.

Rare Earth: Why Complex Life is Uncommon in the Universe, Peter D. Ward and Donald Brownlee. Copernicus, 2000.

The Search for Life on Other Planets, Bruce Jakosky. Cambridge University Press, 1998.

Internet resources

NB Web site addresses are subject to change.

National and International Organisations
British National Space Centre (BNSC)
http://www.bnsc.gov.uk/

European Space Agency (ESA)
http://www.esa.int/

National Aeronautical and Space Administration (NASA)
http://www.nasa.gov/

Natural Environmental Research Council (NERC)
http://www.nerc.ac.uk/

The Natural History Museum
http://www.nhm.ac.uk/

Particle Physics & Astronomy Research Council (PPARC)
http://www.pparc.ac.uk/

Astrobiology
Astrobiology web
http://www.astrobiology.com/

Centre for Astrobiology (Spain)
http://www.cab.inta.es/

NASA Astrobiology Institute
http://nai.arc.nasa.gov/

SETI
http://www.seti-inst.edu/

UK Astrobiology Forum
http://ast.star.rl.ac.uk/exobiology/

Images
Anglo Australian Observatory
http://www.aao.gov.au/

Hubble Space Telescope
http://hubble.stsci.edu/

NASA Image Archive
http://photojournal.jpl.nasa.gov/

Planetary Missions

Earth-like planets
Darwin
http://sci.esa.int/home/darwin

Terrestrial Planet Finder
http://tpf.jpl.nasa.gov/

Jupiter
Voyager
http://vraptor.jpl.nasa.gov/voyager

Galileo
http://galileo.jpl.nasa.gov

Mars
Mariner, Viking
http://nssdc.gsfc.nasa.gov/planetary/

Pathfinder, Global Surveyor
http://mars.jpl.nasa.gov/

Mars Express
http://sci.esa.int/marsexpress

Beagle2
http://www.beagle2.com

Saturn (Titan)
Huygens
http://sci.esa.int/huygens

Acknowledgements

Celia Coyne is thanked for editing my deathless prose into something comprehensible. Gratitude is extended to the Particle Physics and Astronomy Research Council (PPARC) for supporting the research activities of the Meteorites and Micrometeorites Programme. Ian and Jack are thanked, once again, for putting up with me.

Picture Credits

AAO/David Malin: p. 7, 8 top, 9 right, 72, 76
AAO/ Royal Observatory: p. 14 right
Alcatel Espace: p. 78
David P. Anderson: p88
BAS: p. 43
Cambridge University Press, The New Solar System, Beatty, Petersen & Chaikin (eds.) 1999: p. 85 (top)
Courtesy of Beagle 2. All rights reserved, p.32
Courtesy of the Bureau of Land Management: p.31
Laurie Campbell: p. 26
Don Cowan: p. 37
Dorling Kindersley: p. 21 bottom
Mike Eaton: p. 8 bottom, 10, 17, 28, 30, 62-63, 83
Encyclopedia Universalis SA: p. 9
ESA: p. 71
ESA/NASA: p. 75
FAO: p. 85 bottom
John G. Francis: p. 38
Gary Hinks: p. 21
Jodrell Bank Observatory: p. 87
Caroline Jones: p. 36, 45
Mercer Design: p. 32, 59
NASA/Goddard Space Flight Center-Scientific Visualization Studio: title page, p. 52, 57, 81 bottom, back cover
NASA/JPL/Caltech: p. 11, 14 left, 16, 21, 47, 50, 54, 55, 60 bottom, 61, 64, 65, 66, 67, 68, 69, 70, 81 top, 89
NASA/JPL/Malin Space Science Systems: p. 56
NASA/JSC; p. 22, 49, 58 right
NASA/Micheal Rich, Kenneth Mighell, James D. Neill (Columbia University) and Wendy Freedman (Carnegie Observatories): p. 74
NASA/Space Telescope Science Institute: p. 5, 20, 77, 80
PPARC/artist impression by David Hardy: p. 79
Perks Willis Design: p. 18, 25
Dr R. Pamela Reid, University of Miami: p. 34 top
D. Roddy courtesy of the Lunar and Planetary Institute: p. 19
Dr Brian Rosen: p. 34 bottom
Dr Mike Searle: p. 24
Peter Stafford c/o The Natural History Museum: p. 39
Virgo Consortium for Cosmological Simulations. Image generated by J.Colberg: p. 77 bottom
Tony Waltham: p. 40
© WHOI, J. Frederick Grassle: p. 41
© WHOI, Dudley Foster: p. 29

AAO, Anglo-Australian Observatory
BAS, British Antarctic Survey
FAO, Food and Agriculture Organization of the United Nations
PPARC, Particle Physics and Astronomy Research Council
SETI, Search for Extraterrestrial Intelligence
WHOI, Woods Hole Oceanographic Institute